ALSO BY FRANK CLOSE

Charge

Elusive

Trinity

Theories of Everything

Eclipse

Half-Life

The Infinity Puzzle

Neutrino

Antimatter

Lucifer's Legacy

The Cosmic Onion

DESTROYER OF WORLDS

DESTROYER OF WORLDS

THE DEEP HISTORY OF THE NUCLEAR AGE

FRANK CLOSE

BASIC BOOKS
NEW YORK

Copyright © 2025 by Frank Close
Cover design by Chin-Yee Lai
Cover images © bbymedia.store/Shutterstock.com;
© 80's Child/Shutterstock.com; © MaxyM/Shutterstock.com
Cover copyright © 2025 by Hachette Book Group, Inc.

Hachette Book Group supports the right to free expression and the value of copyright. The purpose of copyright is to encourage writers and artists to produce the creative works that enrich our culture.

The scanning, uploading, and distribution of this book without permission is a theft of the author's intellectual property. If you would like permission to use material from the book (other than for review purposes), please contact permissions@hbgusa.com. Thank you for your support of the author's rights.

Basic Books
Hachette Book Group
1290 Avenue of the Americas, New York, NY 10104
www.basicbooks.com

Printed in the United States of America

First Edition: June 2025

Published by Basic Books, an imprint of Hachette Book Group, Inc. The Basic Books name and logo is a registered trademark of the Hachette Book Group.

The Hachette Speakers Bureau provides a wide range of authors for speaking events. To find out more, go to hachettespeakersbureau.com or email HachetteSpeakers@hbgusa.com.

Basic Books copies may be purchased in bulk for business, educational, or promotional use. For more information, please contact your local bookseller or the Hachette Book Group Special Markets Department at special.markets@hbgusa.com.

The publisher is not responsible for websites (or their content) that are not owned by the publisher.

Print book interior design by Sheryl Kober.

Library of Congress Cataloging-in-Publication Data
Names: Close, F. E. author
Title: Destroyer of worlds : the deep history of the Nuclear age / Frank Close.
Description: First edition. | New York, NY : Basic Books, 2025. |
    Includes bibliographical references and index.
Identifiers: LCCN 2024046838 | ISBN 9781541605893 hardcover |
    ISBN 9781541605916 ebook
Subjects: LCSH: Nuclear physics—Research—History | Nuclear
    engineering—History | Nuclear weapons
Classification: LCC QC773 .C56 2025 | DDC 623.4/511909—dc23/
    eng/20250324
LC record available at https://lccn.loc.gov/2024046838

ISBNs: 9781541605893 (hardcover), 9781541605916 (ebook)

LSC-C

Printing 1, 2025

CONTENTS

Acknowledgements   *vii*

PRELUDE   Trinity 1945   1

PART I   THE NUCLEUS REVEALED: 1895–1913
1. The Third Revolution   11
2. From New Zealand to the World   25
3. Otto Hahn and Lise Meitner   39
4. The Nuclear Atom   53

PART II   THE NUCLEUS EXPLAINED: 1914–1932
5. Rutherford "Splits the Atom"   69
6. The Mystery of Beryllium   81
7. Il Papa   91
8. In Bed for a Fortnight   103
9. Moonshine   113
10. The Magicians   127

INTERLUDE   The Birth of Nuclear Physics: 1933 Solvay Conference   141

PART III   RELEASING THE NUCLEAR GENIE: 1933–1939
11. Fermi Explains Beta Radioactivity   149
12. Third Time Lucky   161
13. To Uranium and Beyond   167
14. Majorana's Vision   183

Contents

- 15 A Walk in the Woods   191
- 16 Chain Reaction   207

PART IV NUCLEAR SECRETS: 1940–1960
- 17 "Extremely Powerful Bombs"   223
- 18 A Nuclear Engine   239
- 19 Destroyers of Worlds   261
- 20 The Ulam–Teller Invention   273
- 21 The MADness of Tsar Bomba   293

POSTSCRIPT A Nobel Trinity: Hahn, Rotblat, and Sakharov   305
AFTERWORD 311

Bibliography   317
Notes   321
Index   333

ACKNOWLEDGEMENTS

One afternoon in 2022, I was walking into town with my ten-year-old grandson, Jack, when he started asking—and remarkably telling me what he knew—about Tsar Bomba, the most powerful bomb ever detonated. I had been thinking about the history of nuclear physics for many years, but it was this conversation that was the final spark for this book. So came about this account of how the discovery of a smudge on a photographic plate in 1896—the first hint of nuclear energy—within just seventy years gave humanity the ability to destroy itself in a blast more powerful than anything since that which killed the dinosaurs sixty-five million years ago.

Another fateful occurrence led me to start writing *Destroyer of Worlds* the following year while being treated for non-Hodgkin's lymphoma. From June 2023, during twenty-one weeks of chemotherapy and three of radiotherapy, I used the enforced withdrawal to complete a first draft. The irony then struck me. Several of the saga's heroic pioneers, such as Marie and Pierre Curie, their daughter Irène, her husband Frédéric Joliot, Enrico Fermi, and others, all suffered radiation damage and even fatal cancers because of their research. Today the biological hazards of untargeted radiation are understood.

Radiation's potential for good is also recognised. Radiation therapy harnesses the destructiveness of radioactivity by directing high levels

Acknowledgements

of radiation at cancerous tissues. Another way to use radioactivity in medicine, in much smaller doses, is to label substances by attaching radioactive tracers. The tracers are tiny amounts of radioactive elements, which are injected into the body. Atoms containing these radioactive nuclei take part in biochemical reactions in the same way that the normal atoms do, and by following their radioactive emissions a picture of what is happening inside the patient is built.

Properly directed these radiations succeed, whereas when used at random they can be destroyers. In 2023 I benefited from the positive uses of targeted atomic energy, such as X-rays, gamma rays, radioactive isotopes, and positron emission. The gestation of *Destroyer of Worlds* is therefore very much thanks to Dr Graham Collins and the teams of radiologists, oncologists, and haematologists at Oxford, who are collectively my Destroyer of Cancers. I also owe special thanks to my wife Gillian and our family for supporting me through that time, to many who over the decades helped shape the raw material for this book, and to my agent Patrick Walsh, along with my superb editors, T.J. Kelleher at Basic and Stuart Proffitt at Allen Lane, who helped bring this project to completion.

PRELUDE
Trinity 1945

On the evening of 15 July 1945, a fleet of military buses set out from the nuclear laboratory at Los Alamos near Santa Fe in northern New Mexico. They were filled with four hundred scientists and technicians who for five years had been in a desperate race to make an atomic bomb before the Nazis or the Japanese. They were headed 250 miles south, to the Jornada del Muerto desert, where, under the cloak of darkness, they hoped to realise the fruits of their work.

Shortly after 2:00 a.m. the convoy reached its destination and disgorged its passengers, by now cold and stiff but also in a state of high excitement as they prepared for the first ever man-made nuclear explosion. In the darkness they could discern the faint glow of floodlights some 20 miles distant across the valley floor. Through binoculars the illumination revealed a 30-metre-tall tower, which in outline looked like a lone abandoned oil derrick made of steel. Suspended near its top was a bulbous container, which housed the bomb.

Up until this moment in history, explosions had been chemical in nature, ignited by a spark. The largest planned one had involved 150 tonnes of explosives detonated in 1885 by the US Army Corps of Engineers. The blast, which destroyed an entire island in the East River of New York City and cleared the way for shipping, was heard 50 miles away in Princeton, New Jersey. The calculations of the Los Alamos scientists predicted that the power of a nuclear explosion coming from a mere 6 kilogrammes of the element plutonium would be equivalent to more than 20,000 tonnes of dynamite—the weight of a battleship made of nitroglycerine. This would be in a different league entirely. If the test was successful, it would mean that a destructive power equivalent to five times the entire load dropped over Dresden in 1943, which had taken three nights to accomplish and involved fourteen hundred heavy bombers, could in the future be delivered by a single plane carrying just one bomb.

At least that is what the theory implied; only experiment would show if this really worked. And there was much concern as to whether the calculations were to be believed. The temperature in the explosion would be tens of millions of degrees, far hotter than the heart of the sun, and some feared that this inferno might ignite the atmosphere. The theoretical physicists at Los Alamos had rechecked their calculations and assured the doubters that the atmosphere would survive. Trusting that no mistakes had been made in the arithmetic, the scientists now took their places in the desert and awaited to see what would happen.

At 5100 feet (1565 metres) above sea level, the desert night was cold. There had been storms overnight, but these had moved away. Stars twinkled through a cobweb of misty clouds, but flickers of lightning beyond the surrounding mountains and the occasional sounds of distant thunder still threatened to disrupt the test. The weather forecast predicted that this would clear within a couple of hours, and so the test was given the go-ahead for 5:30 a.m., shortly before dawn.[1]

The canister suspended in the tower contained slices of plutonium, a substance so unstable that it is no longer found naturally on Earth, all primaeval atoms of the element having long since decayed. As in the solar system Pluto lies beyond Uranus, so in the table of atomic elements is plutonium an outlier beyond uranium, the heaviest naturally occurring element. Plutonium, nearly twice as dense as lead, is radioactive, spontaneously converting into uranium and other more stable elements. The plutonium to be used in the bomb test had been created in nuclear reactors by bombarding uranium with neutrons. It had taken months to breed its 6 kilogrammes, one atom at a time.

In small amounts plutonium degrades, releasing energy slowly, but if you bring enough together—surpassing what is known as the *critical mass*—plutonium will blow up of its own accord. This is not an ordinary explosion like TNT or dynamite where the outer reaches of atoms liberate chemical energy. It is the result of a sudden release of nuclear energy, energy that has been locked in the heart of atoms since before the Earth was born.[2]

To exceed the critical mass, the idea was for the pieces of plutonium, which were initially located safely near the surface of the spherical container, to implode, forming a supercritical mass at the centre. To make this happen, conventional explosives surrounded the sphere in the pattern of twelve hexagons and twenty pentagons often used to make footballs.

A small radioactive source inside the device spontaneously released neutrons, electrically neutral constituents of atomic nuclei. When one neutron hits the nucleus of a plutonium atom, it will split that nucleus in two, a process known as *fission*. Plutonium nuclei themselves contain neutrons; fission liberates both energy and two or three of those neutrons. If the plutonium sample is smaller than the critical mass, these neutrons will escape before they can induce further fissions, but for larger volumes above the critical mass, further collisions will take place, inducing more fissions and releases of energy. The first of these liberates energy and three more neutrons, which in their turn can hit further atoms, releasing more energy and a third generation of neutrons. This continues, producing a fourth generation, a fifth, and onwards such that within less than a thousandth of a second there is an exponential growth of neutrons and release of energy. The entirety will explode in less than the blink of an eye.

Shortly after 5:00 a.m. everyone was alerted by announcements on loudspeakers that the test was imminent. The countdown was broadcast. Now was the time to don dark goggles to counter a flash so bright it was expected to penetrate eyelids, plaster on sun cream to protect skin from radiation, and then lie prone in the sand facing away from the blast. A few seconds after 5:29 a.m., electric pulses travelling along miles of cable to the tower reached their destination: the explosive charges on the surface of the sphere. These detonated causing the shell to collapse, imploding the individual pieces of plutonium metal into a concentrated lump greater than the critical mass at the centre of the bomb.

Where moments earlier a radioactive source had produced neutrons which were effectively harmless, now, in the heart of the assembly where the plutonium had imploded, there was no escape for them. The unstoppable inferno began when one neutron set off one fission and immediately spawned a chain reaction, which flashed through the compacted metal faster than lightning. Energy stored within the nuclei of plutonium atoms was released with an explosive power previously unknown on Earth. The force of the blast was billions of times stronger than atmospheric pressure, the temperature four times hotter than the heart of the sun. Nothing on Earth can withstand such conditions.

Miles away across the valley, not everyone had shut their eyes or lay facing away; some took a surreptitious peek at the distant tower. Those who did saw a momentary flash like a distant sunrise but as bright as high noon. Early risers in Tularosa, over the horizon 40 miles to the southeast, saw what appeared to be a premature dawn in the north, far from where a normal sunrise would happen. The blast emitted light over the entire range of the electromagnetic spectrum. The scientists, in groups positioned miles away from the explosion, saw the crevices and peaks of the surrounding mountains briefly illuminated by a strange greenish glow.

In addition to this flash of visible light, the multimillion-degree temperature of the blast radiated heat. This man-made artificial sun vaporised the metal tower and in an instant fused the surrounding desert sand into glass. A buoyant gas of shattered atoms rose rapidly, causing turbulent vortices to curl downwards around its edges. These formed a central column that drew up debris and condensed atmospheric vapour to form the stem of what looked like a gigantic mushroom, the first time such a thing had been seen.

Even after spreading 10 miles, the hotness scorched the watchers' skin, one describing it as like "opening a hot oven door with the sun coming out like a sunrise".[3] The radiant blast also included X-rays and lethal gamma rays, light with higher energy than the visible spectrum. These invisible rays passed through the watchers' sunscreen, penetrated their skin, and passed through their bodies. In that moment the intense flux broke strands of their DNA, causing genetic damage that years later would lead to catastrophic mutations and, in some cases, life-threatening cancers.

In many filmed renditions or newsreels of atomic blasts, the visual drama gains added intensity thanks to the simultaneous accompaniment of the deafening sounds of the explosion.[4] That is not what the watchers lying 10 miles away experienced, however. They saw portents of the apocalypse, but in total silence.

This is because, whereas radiant heat, gamma rays, and the vision itself all travel at the speed of light and reached the observers in a mere instant, sound advances only 330 metres every second or a mile in about five seconds. So, for up to a minute the desert night remained quiet but for the cheers of the scientists. The hurrahs died as everyone took cover, the awesome vision of the cloud of radioactive debris already rising miles into the stratosphere forewarning them of what would arrive within less than a minute.

Watchers 10 miles away had envied those in the relative front row, some 6 miles from ground zero, but now the vision showed the blast to have been bigger than most had anticipated, and even 10 miles felt too close for comfort. A bang loud enough to shatter the eardrums was on its way, shortly to be followed by a tsunami of high-pressure air, the mechanical effect of the explosion, moving at about 500 miles an hour, blasting everything in its path with irresistible force.

Lying flat on the ground, heads away from the explosion, faces down, and hands over their ears, they felt the storm pass with a deafening boom. Debris flew past. The bang echoed from the valley walls, rolling back and forth, as for several minutes in a sonic form of radar the shock waves mapped mountain ranges up to 50 miles away. Meanwhile the mushroom cloud rose higher, its colour changing until its light subsided and, in the east, the red glow of the real dawn took over.

The nuclear age had arrived. It was 05:29 a.m. in New Mexico on 16 July 1945.

It's little appreciated that there were three Industrial Revolutions.

The First Industrial Revolution was powered in the eighteenth century by the engineering genius of James Watt, the Scottish inventor who radically improved the power and efficiency of steam engines. The dynamo of the Second, which began in the nineteenth century, was Michael Faraday's discovery at London's Royal Institution of electromagnetic induction, a fundamental principle of electric motors. And a Third Revolution took place in the first half of the twentieth century when we discovered how to release the vast reserves of energy locked within the nuclei of atoms.[5]

Steam power involves water changing from liquid to vapour; the mutual cohesion between water molecules—H_2O—suddenly disappears at 100 degrees Celsius, but the individual atoms of hydrogen and oxygen in those molecules are unchanged. The first direct hints of activity within atoms themselves came at the cusp of the twentieth century when the detection of radioactivity and the atom's constituent parts revealed the presence of a vast reservoir of atomic energy.

In 1897, the discovery of the electron identified the carrier of electricity, the source of magnetism, and the engine of the Second Revolution. Electrons are electrically charged particles that exist in the outer reaches of atoms. Electric current is the flow of electrons—through wires, liquids, and gases—which involves the disruption of atoms as they give up or receive electrons and reorganise themselves to enable free passage of the current. Chemistry is a result of electrons moving between atoms, reconfiguring their molecular combinations. Throughout history to this juncture, and well into the twentieth century, atoms and their electrons were the engines of science and industry. Yet in all of eighteenth- and nineteenth-century technology, the atomic nucleus was passive, a static lump of positive charge, the seed around which electrons whirl to build atoms.

There would have been no nuclear age had we not first discovered this hidden jewel and found that it too has an internal structure comprised of protons—massive positively charged particles—and their near-twins, neutrons—massive electrically neutral particles. Reorganise those constituents and energy can be released in amounts that are over a million times larger—atom for atom—than anything made available because of the first two Industrial Revolutions. This discovery, which marked the dawn of the Nuclear Age, coincided almost exactly with the start of the Second World War. Its empirical validation heralded the end of that conflict.

Of all the bricks in nature's construction kit, the nucleus is the most deeply hidden. In our daily lives its only visible presence is the sun, a nuclear furnace converting six hundred million tonnes of hydrogen into helium every second.

The temperature in the sun's centre where this alchemy takes place is about fifteen million degrees Celsius. Heat the environment to tens of millions of degrees, or focus the equivalent amount of energy, and the nucleus can be revealed. In Earth's ambient conditions, however, nuclei normally stay at atoms' length, cloaked by a net of protective electrons. These electrons are the agents of chemistry, biology, and life, whereas the nucleus sits inert at the centre of their activity, occupying less than a trillionth of each atom's volume. The effects of electrons are visible in electric sparks, lightning, and aurora, whereas radioactivity—the only natural output of the nucleus and the clue to its existence—remains hidden to normal senses. But for serendipity and the insights of genius, the atomic nucleus might have stayed long shrouded from us.

This first inkling of nuclear energy was so trifling that it was almost missed. Instead, the chance discovery in 1896 of faint smudges on a photographic plate in a closed unilluminated drawer inspired a quest to tap and control this new force of nature. Pursuit of this hidden power source began innocently and collaboratively only to be overtaken by world events in the 1930s as the spectre of fascism loomed. In exactly fifty years science solved how to liberate nuclear energy, delivering it in a steady stream as in a nuclear reactor, in the explosive blast of an atomic bomb, or in a "backyard" thermonuclear weapon so powerful that there would be no need to move it from the construction site—as it could destroy all life on earth from anywhere.

For millennia nature had hidden the presence of the atomic nucleus from sight. The clues however were there, and it was in Germany, late one afternoon on a dank November day in 1895, that the saga began.

PART I
THE NUCLEUS REVEALED
1895–1913

1

The Third Revolution

By the mid-nineteenth century, science was beginning to make sense of the material world. The First Industrial Revolution, associated with steam engines, the development of thermodynamics, and the application of Newton's laws of motion, was a century old. Solids, liquids, and gases, such as ice, water, and steam, were understood as made of many microscopic particles—atoms or molecules—in constant rapid motion. The scientists of the day agreed that the average kinetic energy of these particles determines the temperature of the ensemble, their increasing agitation as temperature rises first breaking the frozen grip of ice and ultimately the more fluid bonds of water to liberate the molecules of steam. This *kinetic theory of heat* was a physical link between thermodynamics and the atomic world.

Chemistry was an established science with atoms as its foundation. All atoms of a given element were believed to be identical, indestructible, and impenetrable, something like miniature billiard balls. Individual atoms of one or more elements join to form molecules. These chemical combinations enabled scientists to determine the atomic masses of different elements relative to that of hydrogen, the lightest, at 1 atomic mass unit (AMU). For example, a molecule of water, H_2O, weighing in at about 18 AMU shows oxygen to have an atomic mass of 16 AMU, and carbon dioxide and nitrogen compounds then determine carbon's and nitrogen's magnitudes to be 12 and 14 AMU, respectively.[1]

In 1869, when the Russian Dmitri Mendeleev listed the known elements in the order of their atomic masses, he noticed that similar chemical properties appeared with periodic regularity. This led to his *periodic table* in which he placed elements with similar properties in columns, with the lightest at the top and heaviest at the bottom, and then placed columns side by side in sequence of their relative masses.[2] This alignment created a gallery of elements arranged like an advent calendar, but with gaps. Mendeleev predicted these vacancies would be filled by yet undiscovered elements, a vision that was dramatically confirmed by the discoveries between 1875 and 1886 of gallium, germanium, and scandium in, respectively, France, Germany, and Sweden. At the time of Mendeleev's periodic table, only sixty-two elements were known, the two heaviest being thorium and uranium with atomic weights then determined as 231 and 240 AMU, respectively.[3] Sufficient of their chemical properties were established for Mendeleev to place uranium in a column separated by a gap from that containing thorium. Mendeleev's scheme thereby implied there exists an element between them, which he named eka-tantalum, meaning one place below the element tantalum in the intervening column.

Atoms were the basic bricks of the substances that powered the First Industrial Revolution. Electric and magnetic forces, the cement that builds material structures from those bricks, were heralds of the Second.

The pulleys and automation of the mills, the first transatlantic steamship, and the construction of railways celebrated a mechanistic perspective of nature, so it was natural that scientists visualised light waves too as a mechanical process involving an intangible *ether* through which the light propagates—after all, mechanical things need something to mechanise the motion. Following James Clark Maxwell's theory of electromagnetism in 1865, however, and Heinrich Hertz's confirmation of its prediction of radio waves, even die-hard mechanists agreed this couldn't be correct: Maxwell's theory implies that light is the result of oscillating electric and magnetic fields in empty space, which can transmit heat energy without need of intervening particles.

A book from that era proudly promised to explain the new wisdom.[4] Comprising 1,258 questions, all answered in 274 pages, it began with "What is light?", "What is heat?", and "What are the attributes of heat?" Then came question number four, which highlighted the enigma at the very foundation of the new electromagnetic revolution: "What is electricity?"

Its answer: "Electricity is a property of force which resides in all matter, and which constantly seeks to establish an equilibrium. What electricity really is has not yet been discovered." In the final quarter of the nineteenth century, scientists sought the answer.

Not only does electricity flow along metal wires like a fluid, flashes of lightning show that it can also pass through the air. This inspired the idea that the flow of electric current might be revealed "out in the open", away from the leads that usually hide it. Electricity can also pass through a liquid containing ions—what today we recognise as atoms that have gained or lost some amount of electric charge.[5] When a current passes through acidified water, oxygen forms at the positive terminal—the *anode*—and hydrogen at the negative *cathode*. Known as electrolysis, this phenomenon always produces the same amount of hydrogen gas for a given amount of electric charge that has flowed. The hydrogen atoms become positively charged—ionised—and, as opposite charges attract, are drawn to the negative cathode. It appears that each hydrogen ion carries a fixed tiny amount of positive electric charge.

So much for electricity passing through liquids; would it be possible given the technology of the late nineteenth century to investigate electric currents passing through a gas? One product of the First Industrial Revolution was the vacuum pump, capable of reducing gas pressure inside a tube to less than one thousandth of an atmosphere. When a high electric voltage was applied to two metal electrodes inside the vacuum tube, the rarefied gas conducted electricity and produced its first surprise: eerie coloured glows that shimmered like moonlight within the airless container.

UNEARTHLY VISIONS

A pioneer of this work was the British scientist William Crookes. The ghostly glistening is unearthly even when you know what it is, and to Victorian scientists, working in the dark in all senses of the phrase, it could be unnerving. Crookes had become involved with spiritualism following the death of his brother. Seeking scientific proof of the soul, he became obsessed with the subtle lights in his tubes. Convinced that during seances he had seen "luminous green clouds" and that the lights in his vacuum tube were the same as these phantoms, he announced in 1874 that he had produced ectoplasm.[6]

Although this led to some ridicule, his research revealed the dramatic way that the lights changed as the pressure dropped. At the lowest pressures then possible, the gas discharge broke up into striations—luminous regions separated by blackness—until eventually at very low pressure the gap expanded, making the whole tube between the negative and positive plates a dark space. Yet curiously it still conducted electricity.

Crookes noticed that the glass at the far end of the tube glowed brightly. Clearly, invisible rays must have travelled through the full length of the dark tube, from its negatively charged cathode, and hit the glass at the far end. To check if this was true, he put fluorescent materials in the path of these *cathode rays*, which lit up when the rays hit and enabled their paths to be "seen". The final proof of the rays' reality was that when he placed a piece of metal in the tube, its shadow appeared in the glow at the far end. As to what the rays consisted of, however, no one knew.

In Germany, at the University of Wurzburg, fifty-year-old Wilhelm Röntgen hoped to find out. During November 1895 he was doing similar experiments when by chance he saw an apparition so awful that he wondered if he had taken leave of his senses. By being well prepared, and noticing the unusual, this piece of fortune led him to one of the great breakthroughs in science.

It was approaching midnight on 8 November. Earlier that day, as the wintry dusk was darkening the laboratory, Röntgen had noticed that whenever he made sparks in the tube, a fluorescent screen at the far end of the laboratory appeared to glow slightly. This proved that invisible rays were indeed being produced in the tube and were passing through the glass, crossing the room, and striking the screen, which produced the faint glimmer. After a late meal Röntgen returned to the laboratory. It was now night, but Röntgen closed the curtains to maintain warmth and to ensure the darkness was total. In the blackness the tantalising glow was easier to see. That was when he had a surprise.

He had been tracking the cathode rays by putting pieces of card in their way and noting their shadows, but the remote screen continued to glow whether the cards were there or not as if the rays were able to pass clean through them. He tried to block them with metal, but thin pieces of copper and aluminium were as transparent as the card had been. Somehow Röntgen's electrical device was producing some novel variety of rays able to pass through objects opaque to light. Whatever these were they could not be cathode rays, which as Crookes had already found and Röntgen confirmed cast shadows of any intervening material. At last Röntgen found something to stop them: a small sheet of lead left a shadow, proving that the mystery rays were real.

He moved the piece of lead near to the fluorescent screen and watched its shadow become sharper. Then he dropped it in surprise: on the screen he had seen the silhouette of the metal apparently held by the hand of a dead man. Astonished, he looked at the dark skeletal pattern of the bones of his hand. Doubting what he saw, he took some photographic film for a permanent record. Röntgen had made one of the most momentous discoveries in the history of science, *X-rays*, and had seen for the first time images that are today common in every hospital.

Six weeks later, on the Sunday before Christmas, he invited his wife Bertha into the laboratory and took a shadow graph of the bones of her hand with her wedding ring clearly visible. This became one of the most famous images in photographic history. Within two weeks it had made him an international celebrity. The medical implications were immediately realised, and the first images of fractured bones were being made by January 1896.

FEBRUARY FOG

The X symbolised that no one knew what these rays were, and Röntgen appears to have shown little interest in finding out. But on 20 January 1896 Henri Becquerel, a forty-four-year-old Parisian scientist with a strong record of research into phosphorescence, uranium compounds, and photography, learned of Röntgen's X-rays at a meeting of the French Académie des Sciences. Two doctors showed a picture of the bones of a human hand, which aroused immense excitement, and a copy of Röntgen's paper on the subject was read.[7] Almost immediately, Becquerel wondered if the rays might be related to the natural phenomenon of phosphorescence where some minerals glow in the dark after first being exposed to light. The question at hand was whether phosphorescent light is entirely stopped by opaque objects or consists of invisible penetrating rays like X-rays?

At the Académie's next meeting, on 24 February, only five weeks after having first learned of X-rays, Becquerel reported on his first experiments. He told the assembled academicians how, in his first trials, he had exposed some phosphorescent crystals of uranium salts to sunlight for several hours so that they were energised. After wrapping plates of photographic emulsion in opaque paper and placing the crystals on top, he put them next to one another in a dark drawer. Between the crystals and the photographic plate, he placed an aluminium medallion stamped with the head of a figure in relief. When he developed the plates, he found they had indeed been exposed and, most importantly, contained silhouettes of the medallion. The area under the thinner portions of the medal were darker than under the thicker, which caused the head to be clearly visible in the photograph (Figure 1).

He had proved without a doubt that the crystals were responsible. He also remarked that uranium compounds were particularly good for this. But he was wrong to believe that it was exposure of the uranium to sunlight that provided the energy setting the process in motion. The true secret—that uranium radiates energy spontaneously without need of prior stimulation—was still to be revealed. Like a latter-day Columbus, Becquerel had set off with a wrong hypothesis, which but for serendipity might have been the end of the story.

Seeking further confirmation of what he had found, he planned to continue his experiments, but the end of February in Paris was overcast. Wrongly thinking he couldn't do the experiment without strong sunlight, he put the uranium crystals, photographic plates, and a copper Maltese cross in a drawer and waited for better weather. A succession of grey days left him frustrated and on 1 March he decided to develop the plates anyway. Expecting to see only a weak image at best, Becquerel was astonished once again to find remarkably clear shapes on the film: the outline of the copper surrounded by a foggy smudge (Figure 2). During the following weeks he conducted further tests to confirm that sunlight was indeed unnecessary, but he was obviously already certain. He told the Académie the very next day, 2 March, of his new discovery: activation of uranium compounds can take place in the dark! Becquerel

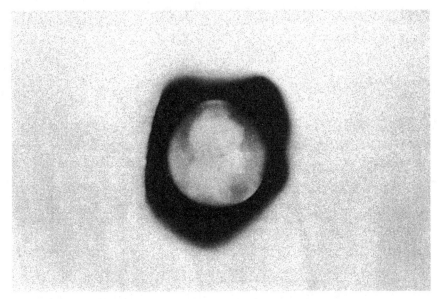

Figure 1. The result of Becquerel's experiment. (H. Becquerel, *Recherche sur une propriété nouvelle de la matière*, Paris: Firmin Didot, 1903; plate 1, No. 2. Wellcome Library, London CC BY-NC 4.0.)

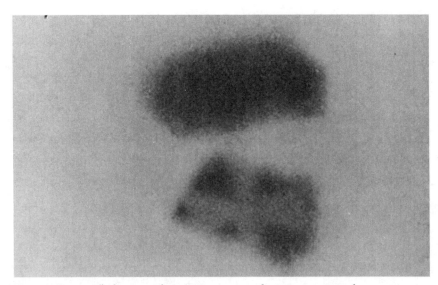

Figure 2. Becquerel's discovery that uranium emits radiation spontaneously.

had stumbled upon the phenomenon known as *radioactivity*—the spontaneous emission of energy by the nuclei of atoms.

Although Röntgen's discovery of X-rays on 8 November 1895 is traditionally regarded as heralding the nuclear revolution, we now know X-rays are beams of very high energy light produced in the periphery of atoms, their only link to the atomic nucleus being that they stimulated Becquerel to take the next step. And, as we now also know, Becquerel moved things forwards for the wrong reasons while fortunately using uranium in the dank days of February.

Becquerel had discovered the first evidence for the release of nuclear energy, but whereas today this is recognised as seminal, it is ironic that at the time it made no special impact. Radiations were the novelty of the decade: cathode rays, X-rays, radio waves, along with light emitted as phosphorescence or even from living creatures such as fireflies were all vying for attention, and Becquerel's rays were originally regarded as merely one more for the list. The real birth of the radioactive era was when the Curies discovered the phenomenon in other elements, in particular radium—so powerful that it glows in the dark. It was Marie Curie who invented the term *radioactivity* and it's with the Curies that the story of radioactivity and its full implications really begins.

FROM POLAND TO POLONIUM

Marie Skłodowska was born in Warsaw in 1867, the fifth and youngest child of poor schoolteachers. Her father, who taught science and mathematics, valued a love of learning more than anything else in the world. He used every opportunity to interest his children in natural phenomena, such as at sunset taking a few minutes to explain the Earth's rotation. Whatever was going on, he would always impart his own knowledge about scientific matters to them.

Marie was the brightest of the five, always the top of her class at school. Women were not allowed at the university in Warsaw, so her father encouraged Marie and her sister Bronya to join with a circle of friends attending the so-called Floating University. The university's faculty and students met secretly by night at different locations to evade the Russian ruling authorities. Education under such circumstances was all but impossible, so in 1891 Marie left Poland, almost penniless, and moved to a more enlightened France. She enrolled at the Sorbonne to study physics and maths, and it was here that she met physicist Pierre Curie, eight years her senior and already on the faculty of the École de Physique et de Chemie Industrielle.

Marie and Pierre married in 1894. Their daughter Irène, a future Nobel laureate, was born in September 1897, following which Marie began her PhD project. Pierre suggested that she investigate the new Becquerel radiation.

Pierre was an expert on *piezoelectricity*, the ability of asymmetric crystals to become electrically polarised when subject to pressure. He suggested that Marie use a piece of piezoelectric quartz to make precise measurements of the pressure exerted by the radiation. This proved key, as she was able to quantify its intensity much more accurately than Becquerel had done. She discovered that the radiation's strength was proportional to the amount of uranium in whatever compound she was using.

Becquerel had suspected the radiation was linked to uranium; Marie had now confirmed it. Whereas Becquerel had concentrated on uranium in the hope of learning more about the radiation, Marie took off in a different direction. Her goal: to see if any other elements showed the phenomenon. She soon found that thorium—a silvery white metal found in granite, where it is more abundant than uranium—also does. This proved the mystery radiation to be a more general natural phenomenon, not a mere curiosity of uranium.

This was the moment when she made her inspired leap. Instead of continuing to examine individual elements, she turned her attention to natural ores. She confirmed that minerals containing uranium and thorium are radioactive, as of course they should be, but to her surprise noticed that the radioactive intensity of some minerals was much greater than could be accounted for solely by their uranium and thorium contents. Most noticeably pitchblende—a brownish-black rock that is largely uranium dioxide—turned out to be very radioactive when dug from the ground. Marie's careful measurements soon convinced her that its radioactivity exceeded that from uranium alone and that pitchblende must contain some additional impurity that is highly radioactive.

The challenge now was to extract this mystery ingredient. The only thing known about it was that it must be radioactive. This led the Curies to work together, developing a totally new science known as *radiochemistry*—the chemical study of radioactivity. Their strategy was to take the ore, dissolve it if possible, and separate its components by standard chemical analysis, and then see where the radioactivity ended up. By repeatedly selecting the highly radioactive extract, the concentration of whatever was causing the radiation was increased. Marie and Pierre found the culprit remarkably quickly. They began the search at the end of 1897; by April 1898 Marie had isolated the source of the radioactivity and by July had determined it to be a previously unknown element. In honour of her birthplace, she named it polonium.

The discovery of polonium was just the beginning, as soon came a more dramatic revelation. Marie continued the purification process and by September the Curies found a further hitherto unknown element whose radioactivity is so powerful that in pure form it glows in the dark and it is warm to the touch: radium.

Today the rays from radium are best known as a treatment for cancer, but when undirected they can cause great damage and suffering. Pierre was killed in a tragic accident in 1906 at the age of forty-six, but his finger joints were already exhibiting swelling caused by radiation, while Marie began to suffer from strange illnesses. Though she survived to the age of sixty-seven, her hands were wrapped to protect the blistering and she eventually died of aplastic anaemia, a condition produced by overexposure to radiation. Her experimental notebooks and even her cookery books were still radioactive fifty years later.

The Curies shared the 1903 Nobel Prize with Becquerel. It was he who had discovered the phenomenon, but it was the Curies who realised its awesome potential. Today, *Curie* is the name for the scientific unit that quantifies radioactivity's intensity.

WHAT IS ELECTRICITY?

In December 1884, just four days after his twenty-eighth birthday, Joseph John ("J. J.") Thomson was appointed the head of the Cavendish physics laboratory at the University of Cambridge. As an undergraduate he had been a star mathematician narrowly beaten into second place in the University's highly prestigious mathematical degree. As a result of this perceived "failure" he decided to diversify into both experimental and mathematical issues in electromagnetic theory.

The appointment was remarkable as well as prescient. The first choice had been Lord Kelvin, after whom the scale of temperature is today named. Universally acknowledged as the leading experimentalist of the nation and as a father of thermodynamics in the mid-nineteenth century who had brought scientific analysis to the First Industrial Revolution, Kelvin preferred to stay in Glasgow. Thomson was a surprising alternative, though it is hard to imagine Kelvin could have achieved greater things than Thomson during his subsequent thirty-five years at the helm. In 1897 Thomson would complete the fundamental understanding underpinning the Second Industrial Revolution when he answered the question: what is electricity?

As cathode rays deposit electric charge where they hit a surface, such as the glass of a vacuum tube, Thomson reasoned they must consist of charged particles, which meant they could be deflected by powerful electric or magnetic fields. When he inserted two metal plates into a tube, one charged positive and the other negative, Thomson discovered that cathode rays were repelled by the negative electric plates and attracted by the positive ones. Like charges repel and unlike attract is the rule, from which he proved conclusively the constituents of cathode rays are negatively charged. Thomson's advantage was in having access to superb vacuum pumps and to more intense electric fields than others at that time. The low pressures achieved with these pumps enabled the charged particles to flow more easily, while the strong electric fields deflected the beams more.

Thomson's key breakthrough was to use both electric and magnetic fields to move the beam of cathode rays around. Upon hitting the glass at the end of the tube, the beam made a small green spot. By surrounding the tube with coils of wire, he created a magnetic field which deflected the beam but in a different way to an electric field. Electrostatic forces deal with electric charges while magnetic forces are concerned with moving electricity; by comparing the effect of both you can calculate the velocity and mass per unit electric charge of whatever is moving. By this means Thomson deduced the properties of the cathode rays' constituents.

He performed a series of experiments using a variety of gases in the tube, different metals in the cathode, and a range of velocities for the cathode rays. Each and every time he found that this ratio—the mass per unit charge—was the same within a factor of about two and, moreover, a thousand or more times larger than anything that had been previously measured in the case of atomic ions. This convinced him that his result was a property of the rays and independent of the gas or cathode materials. If the charged particles were the same as what was responsible for giving charge to ionised atoms, then their masses must be at least one thousand times smaller than any known atom. And atoms were, supposedly, the smallest things of all.

Thomson now made his seminal leap, describing cathode rays as "matter in a new state ... from which all the chemical elements are built up".[8] He named the charged particles *electrons*. The enormous magnitude of the charge-to-mass ratio is because the mass of an electron is but a trifling part of that of an atom—about one part in two thousand in the lightest atom, hydrogen, for example. The electron is a fundamental constituent building block of the atom. When liberated from within atoms, by heat or other forms of energy, the flow of electrons is what constitutes electric current.

Thomson won the Nobel Prize for his work on the conduction of electricity through gases, in 1906. While rightly famed for his discovery of the electron, the first direct evidence of a subatomic world, his greatest legacy was in the inspired guidance he gave as head of the Cavendish Laboratory to a young student from New Zealand. The Cavendish, having in Lord Kelvin missed the father of thermodynamics and appointed in Thomson the scientist who completed the electrical revolution, now welcomed the architect of the nuclear age: Ernest Rutherford.

2
From New Zealand to the World

In 1895, the winner of a British Commonwealth Scholarship to Cambridge University declined the award in order to get married. The runner up, twenty-four-year-old Ernest Rutherford, was tending the potato patch on his parents' farm in New Zealand when news reached him that he had won the prize by default. He threw down his spade and famously said: "That's the last potato I'll ever dig".[1] Forty-three years later, when he died in 1938, the rustic Kiwi was Lord Rutherford of Nelson, Order of Merit, and a Nobel laureate. His ashes are interred in Westminster Abbey near the remains of Isaac Newton.

A scientific titan who "seemed to know the answer before the experiment was made", even Rutherford's secondary achievements would have brought fame to other talented scientists.[2] Among these accomplishments, he was the first to date the age of the Earth, he was ahead of Guglielmo Marconi in long-range radio wave transmission and briefly the world record holder, he invented methods for detecting ionising radiation, and he predicted the existence of the neutron. Yet these are not what immortalised him. As Darwin is synonymous with evolution, Newton with mechanics, and Einstein with relativity, so is Rutherford with the atom. Faraday and electricity spawned the Second Industrial Revolution; Rutherford was father to the Third.[3]

Yet Rutherford was an ordinary boy, the fourth of twelve children. On the family farm his father was a skilled mechanic, an expertise which Ernest inherited in his uncanny ability to devise sensitive apparatus seemingly from whatever came to hand; his mother was a teacher dedicated to hard work, an attribute that marked him throughout his life. Rutherford's birth certificate from Waimea South erroneously recorded his name as "Earnest"—but presciently, for he was gifted with great concentration and perseverance. A colleague years later compared Rutherford's mind to the bow of a battleship: "There is so much weight behind it, it had no need to be as sharp as a razor". Later, he developed a remarkable ability to visualise what was going on at the heart of his investigations, living in the world of the atom "as completely as an historian with the people whose life and times he uncovers".[4]

He was also born lucky. First, that he was not the eldest son, for in rural New Zealand it was preordained that he would have joined his father to run the family business: Rutherford and Son. Second, that he was not a girl, for daughters were expected to help their mother bring up the other siblings. He was academically fortunate too, as every major educational scholarship that he won in New Zealand came only at the second attempt, and but for those financial awards he would never have entered high school at Nelson College nor the University of Canterbury. And as we have seen, the scholarship that took him to Cambridge, igniting his stellar career, was itself a matter of luck.

The scholarship in question was the "1851 Exhibition", a postgraduate award to be held anywhere in the world for two years by a student doing outstanding research in a field of importance to their national industries. In 1885 Heinrich Hertz in Karlsruhe had discovered radio waves—electromagnetic waves oscillating at a frequency of about fifty million times a second, ten thousand times slower than visible light. Rutherford had become interested in the possibility of using these waves for electromagnetic signalling, and his main interest was the magnetisation of iron by high-frequency radiation. This was a subject at the frontier of electrical science and of importance to electrical industries where iron-cored transformers were the hi-tech of the day. In England, J. J. Thomson had suggested a qualitative way to study this, but Rutherford went further: he would design a means of measuring magnitudes. This typified his style: where others used qualitative methods, Rutherford devised ways to quantify measurements.

This was five-star work, worthy of a mature scientist; for a student it was outstanding, yet the scholarship committee only deemed it good enough to be positioned second. Indeed, it might be more accurate to say they placed him last as there were only two applicants; the other was James Maclaurin, a government analyst. Maclaurin had developed ways of extracting gold from quartz, which in 1890s New Zealand had immediate industrial interest and swayed the committee. However, the fine print of the award stipulated that it was only for full-time research and the recipient could not accept other remuneration. The value of the scholarship was £150 whereas Maclaurin's position was already paying him £180. Facing the costs of marriage, he had no wish to give up the money and withdrew.

On the night of Thursday 1 August 1895, Rutherford left New Zealand by steamer, accompanied by his experimental equipment, to pursue studies at Cambridge under Thomson. A biography by New Zealand scientist historian John Campbell wistfully recorded the significance of that juncture: "From then on, Ernest Rutherford belonged to the World".[5]

FROM RADIO TO RADIOACTIVITY

When Rutherford arrived in Cambridge, Thomson was already deep into the research that would lead him to discover the electron. Rutherford came with an invention and a research plan which fitted well with the laboratory's interests. Thomson quickly recognised the student's innate experimental skill and the goal that he was chasing: to produce and detect radio waves over large distances. Hertz had produced radio waves and successfully detected them in the laboratory over a range of about 2 metres; by December 1895 Rutherford was already able to transmit and receive radio over 20 metres and noticed that the waves would pass through walls. For the first time solid matter began to appear not so solid! Early in 1896 he was transmitting over half a mile and designing a device capable of communicating over 10 miles.

Meanwhile the discovery of X-rays and Becquerel's radiation seized everyone's attention. Thomson saw these as the great scientific opportunities of the new frontier. The young Rutherford's work on Hertzian waves had made him noticed by many senior members of the university but pursuit of radio as a commercial enterprise could both interfere with Thomson's focus on the novel radiations and cost the laboratory considerable investment. The development costs of Rutherford's research into radio were mounting. Thomson consulted business contacts in the city of London and Lord Kelvin. If that great scientist had been more enthusiastic, history might have been different, Rutherford becoming famous for radio communications and someone else known as the founder of the atomic age. However, Kelvin's caution led Thomson to decide that Rutherford should instead investigate the new rays; so was the field of radio left open for Marconi to make his name.

X-rays were astonishing for the public, their ability to photograph the inside of things both a novelty and an opportunity. For scientists there remained the question of what X-rays are, of course, but the rays also began to reveal a range of new phenomena, such as their ability to cause air to lose its normal insulating property and become a partial conductor of electricity. This intrigued Thomson who around Easter 1896 invited Rutherford—a student of just two terms standing—to join him in the first studies of how X-rays electrified gases. They deduced that the rays ionise atoms—in other words, give them electric charge. Today we recognise ions to be atoms that have become charged by the removal or addition of one or more electrons. Recall, under the influence of an electric field, ions move to the cathode or the anode depending on the sign of the charge. Thomson and Rutherford proved that both positive and negative ions were produced in the gas, and in parallel to these experiments Thomson noticed for the first time that cathode rays appeared to consist of streams of negatively charged particles.

During these investigations Rutherford designed a very simple instrument to measure both the presence and magnitude of the ionisation. Known as an *electrometer* this device would prove key to much of his research in the next decade. Its construction is an example of Rutherford's genius for using the simplest of tools to lever out the most profound of truths. The basic principle was to measure the deflection of a charged metallic strip—such as gold leaf—in an electric field. If the air around the strip becomes ionised, the charge leaks away—a current flows—and the strip moves. Rutherford timed the movement to measure the rate of leakage and hence the amount of ionisation: the faster the leakage, the more ionisation and the stronger the radiation.

This was pioneering, and precise enough to make a decent quantitative measurement of the radiation's intensity. Initially he had been using radioactivity as a tool to study ionization, but by 1898 his interest had been captured by the radiation itself. The first question: what does it consist of?

Being able to measure the ionisation enabled him to see how the radiation was absorbed by various materials. To do this he covered the uranium source with some aluminium foil. He noticed the intensity of the radiation immediately dropped dramatically, but when he added more pieces of foil, the radiation appeared to maintain its strength. Only when he added more covering did he find that the radiation was indeed diminishing, but very slowly. A paper-thin leaf of aluminium had been sufficient to cut the intensity by a half, but Rutherford discovered that several millimetres were needed to absorb the rest.

He deduced that there were two types of radiation, which he named *alpha* and *beta* after the initial letters of the Greek alphabet. The alpha radiation is very easily absorbed—the thin leaf had been enough to stop it—whereas beta radiation is more penetrating.

Rutherford's discovery that there were different types of radiation was published in 1899 and followed up by scientists in Germany and France as well as at the Cavendish Laboratory itself.[6] Their experiments showed that beta rays are deflected by magnetic fields and consist of streams of negatively charged electric particles like cathode rays, though moving more slowly. Today we know that these particles are electrons, but they are always referred to as beta particles to acknowledge their nuclear origin and distinguish them from their atomic counterparts.[7] However, no one at the time was able to deviate alpha rays from their path. This is because alpha particles are more than seven thousand times heavier than beta particles: the deviation caused by a given amount of force is inversely proportional to the mass of the object, which means that massive alpha particles can be almost impervious to a magnetic force that can nonetheless easily deflect the more insubstantial beta particles.

Rutherford's ability to detect radioactivity soon confirmed that it is not unique to uranium and thorium. Meanwhile, in France, the Curies showed that both radium and polonium atoms can continuously radiate energy, year in, year out, seemingly out of nothing. This observation challenged one of the fundamental principles that flowed from the First Industrial Revolution, namely the first law of thermodynamics: energy cannot be created or destroyed, only transformed. If atoms of elements such as radium and polonium could liberate energy without change, this would be like the discovery of perpetual motion and would violate the foundations of thermodynamics, for nothing goes by itself. By the cusp of the twentieth century it was clear that the phenomenon of radioactivity implied something novel and profound about the nature of atoms.

At the end of 1898, Thomson's brilliant protégé was offered the position of professor of physics at McGill University in Montreal, Canada.[8] The post made minimal demands on teaching and the university physics department had a well-endowed laboratory, which would enable Rutherford to pursue his investigations into radioactivity. It is remarkable to realise that less than three years earlier, when Rutherford arrived in Cambridge, scientists believed atoms to be the fundamental stable seeds of all matter. With his move to McGill, Rutherford began to overturn these beliefs.

As the twentieth century began the questions facing physicists were these: What are X-rays? What is radioactivity? How can endless amounts of energy be radiated in radioactivity while satisfying the constraints of energy conservation and yet apparently without obvious change? And with Thomson having discovered the negatively charged electron and hypothesised that it is a fundamental component of all atomic elements: how is the electronic atom constructed?

"BEYOND ORDINARY CHEMICAL THEORY"

Rutherford's appointment as MacDonald Professor of physics at McGill University at age just twenty-seven was a spectacular recognition of his talent. At this time, he was involved in a manic burst of work and about to cause what has been described as "one of the biggest revolutions in scientific thought".[9] In preparation for this research, he arranged for supplies of uranium and thorium to be sent from Germany to Montreal ready for his arrival.

Rutherford's immediate goal was to determine the energy of the radiation that uranium, thorium, and other radioactive substances emit. Heat is a form of energy. The physics department at McGill had experts in heat transfer and there was a lot of specialised heat measurement equipment available, so Rutherford was well placed to determine the energy of radioactivity. He did so by conducting three separate experiments.

First, he irradiated metal with X-rays, measured the amount of heat this produced, and from this determined the amount of energy in the X-rays. Next, he used X-rays from the same source to ionise gases. As he had previously determined the rays' energy, he was able to deduce the amount of energy involved in the ionisation process. The final part of the trilogy was to ionise those same gases by radioactivity. By back-calculating from the rays' ability to create ionisation, he determined the amount of energy being released through ionisation by the radioactivity. The result was that he knew how efficiently radioactivity caused ionisation and could thereby determine the amount of energy carried by the rays per second—in effect, their power.[10]

The result astonished him. At the time he noted: "If we suppose that the radiation has been going on at its present rate for the course of 10 million years, each gramme of uranium has radiated at least 300 calories".[11] This is actually an underestimate, for Rutherford had only measured alpha radiation and, moreover, we now know that the age of the Earth is of the order of *thousands* of millions of years, not just ten million, and that the uranium had been doing this for all of that time. In any event his conclusion was seismic: he remarked that such a quantity of energy was far greater than could be accounted for by "ordinary chemical theory" and revealed the presence of "agencies at work on the Earth's surface of a far more powerful order" than anything previously known or imagined.[12] This was the first insight of nuclear energy, what is commonly called *atomic energy*.

Why did "300 calories" make Rutherford so certain that this was completely novel? Indeed, what is 300 calories?

A calorie is defined as the energy needed to heat 1 gramme of water by 1 degree Celsius, so 300 calories could transform 3 grammes of melting ice into boiling steam. Meanwhile the 1-gramme ball of uranium capable of doing this would be a mere millimetre across. The ability for something so tiny to produce sufficient energy to boil water was far beyond anything that chemistry could explain, as Rutherford realised.

Another way of expressing this is to recall that the concept of energy emerges from Isaac Newton's laws of motion. Newton supposedly had his insight about gravity when an apple fell from a tree before him. Rutherford's 300 calories correspond to the kinetic energy with which a couple of obese Golden Delicious or four Granny Smith apples would hit the floor had they dropped from the height of the Eiffel Tower. Conversely that is the amount of energy that would be needed to blast them all the way to the top of that giant from the ground. By any measure there are vast amounts of energy hidden within that little droplet of uranium.

"A CAPRICIOUS VARIATION"

Rutherford had established the nature of radioactivity in uranium and at McGill he began to study the phenomenon in the element thorium. To do this he followed the same procedure as he had done in the case of uranium. First, he placed two parallel metal plates near to one another, charged the upper one electrically, and dispersed radioactive material on the lower. The radioactivity ionised the air between the plates, which caused the charge on the upper plate to leak away. An electrometer measured the leakage current.

The radioactivity of thorium persisted like that from uranium, but it also exhibited some unique characteristics. In addition to the alpha and beta rays that Rutherford had found in uranium, he noticed a more penetrating form, today known as gamma rays. Gamma rays consist of very high energy light, much more energetic even than X-rays.[13]

More remarkable though was that the amount of radiation exhibited what he called "a capricious variation".[14]

Rutherford noticed that the intensity was affected by draughts from unlatched windows or when someone opened the lab door. Where most experimenters would say "shut that door" and continue, Rutherford wondered: "Why?" A puff of air hadn't disturbed the powerful rays from uranium, so what was special about thorium?

One of Rutherford's leading biographers describes how what happened next typified Rutherford's style. An anomalous result such as this would worry that "rather slow but very powerful mind until an answer emerges".[15] Then would follow a furious burst of experimental work.

After much thought and further tests, Rutherford found the answer: in addition to emitting radioactivity, thorium was also giving off an "emanation" in the form of a gas, which was itself radioactive. If a puff of air blew some gas away from the metal plates, there would be a momentary drop in the ionisation and the measured intensity would fall. Meanwhile thorium would continue to emit the emanation. Once the air settled again, the gas would remain and the intensity of the radioactivity of both thorium and gas would return to its previous level.

This theory fitted the facts, but Rutherford needed to prove it. To do so he performed a sequence of experiments, working late into the night for several months, every day of the week but Sunday.

He blew air gently between the plates, which sent the emanation up a long tube attached to a bottle containing an electrometer. When the gas arrived in the tube, the current in the electrometer rose and within a few minutes became a constant, which confirmed that the gas itself was radioactive. This was already significant progress, but now came the real surprise. When the airflow stopped, the current decayed even though the gas remained. Rutherford remarked that the emanation "isn't constant but gradually diminishes" in a "geometric progression with time".[16] In about a minute the radiation had fallen by a half, after two minutes to one quarter, and then to an eighth after three minutes. This behaviour is typical of physical processes where the change at any moment is proportional to the amount of the substance available for that change. It is like a COVID pandemic, in reverse. The COVID virus spread exponentially in proportion to the number of people already infected; what Rutherford was observing was an exponential *fall* reflecting that the number of active atoms in the radioactive "emanation" was decreasing. The time for the intensity to fall to half the value is called the *half-life*.

The reason Rutherford had thought the radioactivity of uranium and thorium to be overall constant is because their half-lives are immense.[17] That of uranium is four and a half billion years, almost the same as the age of the Earth itself, while that of thorium is even longer: at fourteen billion years it is comparable to the age of the universe. The half-lives of radioactive elements are like yardsticks of time by which the percentages of different radioactive elements in ores today reveal how long they have been there. The relative abundance of uranium, thorium, and other radioactive elements in a variety of minerals would later become key to ageing the Earth itself. In his first explorations around 1900, however, to all practical purposes the measured intensity would have remained the same over the period that Rutherford was observing in his laboratory. The half-life of the thorium "emanation" we now know is about 54 seconds, enabling him for the first time to see the drop in intensity even as he was making his measurements.

This discovery of half-life would turn out to be key to unravelling the mysteries of radioactivity. As further examples of radioactive elements were found, the first green shoots of uranium, thorium, radium, and polonium rapidly became an impenetrable jungle. Each has a unique half-life, like a fingerprint that can identify the specific source, and these measurements began to bring order to the mystery. In France, Becquerel and the Curies had also noticed that radioactivity in both radium and polonium gradually lessened, but they had drawn no conclusions from it.

THE ALCHEMISTS

Having discovered that thorium emits a radioactive gas, the question was: what is it? To find the answer, Rutherford needed a chemist. He invited his friend Professor James Walker, who had joined McGill the same day as him, but Walker declined, explaining that he was an expert in organic chemistry and what Rutherford needed was someone specialising in gas analysis. Having passed up on Rutherford's offer, Walker's name is missing from the annals of fame. Instead, Rutherford's name would become forever linked with that of Frederick Soddy, a twenty-three-year-old junior assistant in the chemistry department, who had recently arrived from Oxford University and specialised in that very field.

Soddy had been working in the physics laboratory, studying the effect of light on chlorine, and the pair had come to know and respect one another, even though they were opposites in many ways. Soddy, the son of a London merchant, had been educated at an English public school and Oxford University, in stark contrast to Rutherford's country upbringing and distrust of class privilege. Soddy, clever with a quick mind, was articulate and could write about science with ease; Rutherford, more ponderous but blessed with deep penetration and breadth of vision, could appreciate the richer philosophical implications of their work.

Their attitudes to atomic matter differed widely, too. Soddy being a traditional chemist couldn't accept that there is anything smaller than atoms; Rutherford, on the other hand, was convinced by radioactivity and Thomson's discovery of the electron that atoms are not the final frontier. They had even taken opposite sides in a debate at the McGill Physical Society in March 1901, titled "The existence of bodies smaller than an atom", which Rutherford had organised. Now they were to collaborate. Starting from opposite directions they came to a common conclusion that would overthrow each of their conceptions about the nature and immutability of the atomic elements. This all came about when after three months of careful analysis Soddy conclusively identified the gas as a chemically neutral element "of the argon family".[18]

In 1894, a Scottish chemist, William Ramsay, and an English chemist, Lord Rayleigh, had independently discovered a chemically inert gas which they named argon, after the Greek word for "idle". That argon is one of a family of inert gases that form a column in Mendeleev's periodic table of the elements became clear four years later when Ramsay discovered neon, krypton, and xenon. A lighter example—helium—had been known since 1868. The member "of the argon family" that Soddy was now announcing turned out to be a sixth, whose atoms are the heaviest of the sextet. Not until 1910 was he able to identify this sixth inert gas unambiguously as radon and confirm it to be the "argon-like" emanation from thorium.[19]

The two scientists had established radioactivity to be a spontaneous disintegration of the atom, where one kind of matter changes into another having different chemical and radioactive properties, with simultaneous emission of energy in the form of radiation. Their revelation of transmutation of the atomic elements proved directly what the earlier discovery of radioactivity had hinted at: the age-old belief in indestructible unchanging atoms was false. Rutherford famously said to his colleague, "They will accuse us of being alchemists", alluding to the long-discredited seventeenth-century quest for a means to transmute base metals into gold. In detecting the spontaneous transmutation of thorium atoms into radon, Rutherford and Soddy had the equivalent of the Midas touch.

Rutherford's caution was well judged, however. The pair were just twenty-five and thirty-one years old and their claims made some senior colleagues beg them not to publish for fear it would discredit McGill. Others were more open-minded, accepting that the youngsters had made a paradigm shift. Scientists in the United Kingdom agreed as the following year, 1903, they elected Rutherford to Fellowship of the Royal Society, one visionary predicting: "Someday Rutherford's experimental work [will] be rated as the greatest since that of Michael Faraday".[20]

A MAGICAL EVENING

Meanwhile in Paris, Pierre and Marie Curie were working with their newly discovered radium, which was proving to be one hundred thousand times more active even than uranium. It was not until August 1903 that Rutherford and his new wife, Mary, met them for the first time. Rutherford was still based in Canada but was visiting the United Kingdom that summer to chair a discussion at a meeting of the British Association for the Advancement of Science in Southport. The visit to Paris was arranged by Paul Langevin, a French physicist whom Rutherford had known during their time together in Cambridge and who was now a colleague of the Curies. In addition to the pleasure of meeting his fellow explorers, he hoped to obtain some of the radium which they had extracted in their laboratory. Radium's greater power promised to make it a more intense probe than the uranium and thorium to which he was currently restricted.

The group dined at the Langevins' home. After sunset they gathered on its roof garden, where Pierre Curie promised a breathtaking demonstration.

Rutherford recalled that the sky was already dark enough for the first stars to be visible. There was no moon, and in those days little scattered light from Paris itself. Pierre put his hand into his waistcoat pocket, dramatically extracted a small transparent tube and held it between his thumb and forefinger against the starry background. An unearthly bluey-green "bright luminescence" more vivid than the Milky Way illuminated the garden.[21] The power of radioactivity was pouring out of radium "like an Aladdin's lamp" for all to see in the darkness.[22] The glow was also enough to show the inflamed state of Curie's hands, a foretaste of the radiation burns that overexposure to radioactivity could cause.

After this otherworldly moment, they returned indoors where they discussed what they had seen and what their various experiments had taught them. It was only now, after they compared experiences and weighed up Rutherford's measurement of uranium's hidden energy store, which now appeared trifling relative to radium's luminous display, that they fully absorbed the profound implications of their discoveries.

3
Otto Hahn and Lise Meitner

FROZEN ENERGY

Soddy returned to England in 1903. His collaboration with Rutherford had made this once fundamentalist chemist a born-again convert to a new religion: atomic physics. Back in the United Kingdom, Soddy popularised the phenomenon of radioactivity, describing how he and Rutherford had showed that atoms of radium and other radioactive elements could continue to emit heat for thousands of years, implying that they contain vast reserves of energy.

Soddy both lectured and wrote about radioactivity. He had a gift for creating pictures in words and proselytized that "we stand today where primitive man first stood [upon discovering] the energy liberated by fire". He saw the positive opportunities of this new source of energy, describing it as free and effectively perpetual, capable of "transforming the deserts" and making the world "one smiling Garden of Eden".[1]

However, although the amounts of energy were in effect unlimited, the power—the rate of energy release—was too slow to be of practical use. One example touted for the implications of atomic energy was that a few kilogrammes of radium contain enough energy to drive a ship across the Atlantic Ocean. The vast energy from radium had been emitted for thousands of years, and while it was undoubtedly true that you could traverse the oceans if you used the supply for long enough, it would be impossible to do so in any sensible amount of time such as a week. Therein lay the catch. Not just uranium and radium but all radioactive elements appeared to liberate energy at a similar sluggish rate, too slow to drive a turbine or to do much useful. Radioactivity is intrinsic to the nature of the element, impossible to turn on or off, or to alter in any way. Heating, compressing, or mixing radioactive sources together had no effect. The elements continued to drip-feed radioactive energy at a rate determined by some yet unidentified chronometer. Nonetheless, if it was somehow possible to tap the source, the implications could be huge as enormous amounts of energy could be obtained from a small amount of matter. In 1904, having moved to the University of Glasgow, Soddy surmised in a lecture to the Royal Corps of Engineers that if the rate of disintegration could be controlled at will, a fast release of radioactive energy "could be employed as an explosive more powerful than dynamite".[2]

Soddy produced an account of his lectures in a book, *The Interpretation of Radium*, published in 1909, which gripped the attention of novelist and science visionary H. G. Wells. Soddy's conjectures inspired Wells to write *The World Set Free*, a novel in which he envisaged the consequences of exploiting the vast energy store within atoms. He imagined a future where, in 1933, scientists discovered how to make natural chemical elements radioactive and thereby release large amounts of energy. (A foretaste of what would occur, in reality, in

1934.) In Wells's vision, however, the ability to make energy industrially available at negligible cost rendered conventional industrial processes too expensive. Economic chaos ensued, leading to war. As to the new weapons of war, Wells turned to Soddy's speculation about the "explosive more powerful than dynamite". Wells coined a name for this new weapon: *atomic bomb*.

In 1905 a Swiss patent clerk, Albert Einstein, produced his theory of special relativity, which gave a new insight into the relationship between matter and energy. The equation $E = mc^2$ is today so famous that it adorns tee shirts, though its profound implications are less well advertised. One that will be key to much of this saga is that while motion gives rise to kinetic energy, a body with mass m at rest contains an intrinsic amount of energy, quantified by that formula. To the scientists enraptured with the mysteries of radioactivity and transmutation of the elements, this gave a new vision of matter as "frozen" energy, the exchange rate between the atomic mass and energy accounts being given by the speed of light—that's the quantity c—squared. The kinetic energy of an entity of mass m, moving off from rest at speed v, is $(1/2) mv^2$, which is trifling compared to mc^2 as c is some 300,000 kilometres per second. All were immediately struck by the implications: thanks to mc^2, atoms contain vast reservoirs of energy.

Atomic masses were key to understanding what was going on. The change in atomic mass during a radioactive transformation implied a change in the frozen energy of the initial and final atoms. Thanks to Einstein's breakthrough, Rutherford, Soddy, the Curies, and all in the hunt realised this energy must be what was being released in radioactive decay. However, while Einstein's equivalence between mass and energy explained the appearance of that energy, it said nothing about *how* the energy got released. That would be for experiment to determine.

In addition to increasing public awareness of radioactivity and the promise of atomic energy, in his first decade in the United Kingdom Soddy brought order to the confusing proliferation of radioactive elements. He announced that atoms of the same element could emit different amounts of radioactivity, or even none. This overthrew another paradigm in showing that all atoms of a given element are not identical. In other words, atoms that have the same chemical properties can have different properties when it comes to radioactivity, some being highly radioactive while others might be mildly so, or even completely stable, showing none. Soddy introduced a term for this. He said that a chemical element can have different *isotopes*—from the Greek meaning "same place" (in Mendeleev's periodic table of elements).[3] He did this only in 1913, so for a decade there was no map through the thickening jungle of radioactive sources. The misinterpretations of that first decade baffled scientists. Lead players in the discoveries that would bring Soddy to his breakthrough were an unlikely pair: an Austrian Jewish woman, Lise Meitner, and a mustachioed doppelganger for the German Kaiser, named Otto Hahn.

OTTO HAHN

"One fine day in the autumn of 1905", a young German chemist, Otto Hahn, arrived at McGill University to work with Rutherford.[4] A gymnast, skier, and mountain climber, Hahn was born in Frankfurt in 1879. By the turn of the century, he was training as a chemist at the University of Berlin. This was when the Curies' discovery of radium was making radioactivity the exciting frontier in atomic science. That Hahn's life work would be in this field, let alone that he would be at the forefront, was happenstance.

He had been offered a post in the chemical industry but as a prerequisite was required to take a year to gain a better knowledge of English. So he went to London, to University College, to work with Sir William Ramsay, who had just discovered the inert gases. Ramsay asked Hahn if he would like to work on radium. When Hahn said he knew nothing about radium, Ramsay replied that it's an advantage to be able to enter a subject free from preconceptions. He gave Hahn a cup of about 100 grammes of a barium compound, said it contained about 9 milligrammes of radium—that is about one part in ten thousand—and told him to separate it from the barium by a standard chemical method known as fractional crystallisation. The goal was to prepare pure radium compounds to check the element's atomic weight.

What Hahn expected to have been straightforward, however, turned out to show some unexpected inconsistencies. Which is where Hahn's natural curiosity and innate scientific skills came to bear as, like Rutherford, he focused on the anomaly to understand what was responsible. This is what happened.

The chemical separation of radium from the barium should have taken all the radioactivity with it and left the remains without any, but Hahn noticed that the activity of the remnants was quite considerable. Some of this was yet unseparated radium, which he could identify by its four-day half-life, but in addition there was a radioactive residue with a half-life of about a minute, which he identified as thorium. The reason for thorium being there was because Ramsay's sample hadn't been extracted from a pure uranium ore but from thorianite, a mineral from what was then Ceylon—today Sri Lanka—which contains a high percentage of thorium. But the activity was much greater than that of thorium previously known. Hahn concluded that he had discovered a new element, and in recognition of its radioactivity he named it radiothorium.

Ramsay of course was very pleased that following his own discoveries of elements, yet another element had been discovered in his lab. He advised Hahn to give up ideas of an industrial career and to focus instead on research into radium. The Berlin Chemical Institute offered Hahn a post with the prospect of a lectureship later. First, however, he decided to get a more thorough grounding in this new field of radiochemistry and moved to Montreal, to work with Rutherford.

Not everyone was persuaded by Ramsay's enthusiastic promotion of Hahn's new "element". Whereas Ramsay was acknowledged by all to be a great expert in chemistry, his knowledge of *radioactivity* was limited. Rutherford received a sceptical letter about Hahn's claim from Bertram Boltwood, a leading American radiochemist.

Boltwood was someone to be reckoned with. He had worked at Yale University until 1900 when he established a consulting firm of mining engineers and chemists. Having become interested in radiochemistry, by 1904 he showed that many radioactive substances decay into other radioactive elements. For example, he was the first to identify that radioactive decay of uranium produces ultimately lead, this coming about because uranium is in effect at the top of a radioactive ladder, the rungs below it being the elements thorium, radium, radon, and polonium, with successive radioactive transitions moving down the ladder one rung at a time until stable ground—lead—is reached. With lead as the durable end product of uranium's activity, he developed a way of dating rocks from their relative contents of these two elements. Effectively, the amount of uranium that had turned into lead was revealed in the rock: the lower the ratio of uranium to lead, the older the rock. Boltwood's pedigree was impeccable, his opinion stark: "The substance [radiothorium] appears to be a new compound of thorium and stupidity."[5] It would be nearly a decade before Soddy explained what had happened: Hahn had discovered an isotope of thorium.

With Rutherford, Hahn established that the energy emitted in alpha radioactivity is carried by positively charged particles, known as *alpha particles*. They did this by using powerful magnets which deflected the alpha particles enough to show they were positively charged and responded like helium ions. The following year Rutherford confirmed this by allowing alpha particles to pass through the thin walls of an evacuated tube and form a gas. An electric spark then heated the gas, which emitted the spectrum of helium.

Hahn measured the distances that alpha particles would travel from different radioactive sources. Alpha particles are easily absorbed, which is how Rutherford had originally distinguished the short-range alpha radiation from the more penetrating beta rays. The faster the particles are moving at the outset, the further they will reach before coming to a stop. Measurement of their ranges in air or through thin sheets of assorted materials enabled Hahn to determine their relative energies.

He next showed that the alpha particles emitted from thorium are deflected by magnetic fields in the same way as those from radium. Their ranges gave a measure of their relative energies, and from the amount of deflection Hahn established that the particles emitted in both cases have the same mass. Here was the first step in establishing that the particles emitted in alpha radioactivity by different elements are all ions of helium.

Hahn meanwhile discovered another radioactive "element", which he named mesothorium. The more he and Rutherford investigated, the more radioactive so-called "elements" were found. Soon there was a jumble of them, too many to fit into Mendeleev's periodic table of elements. By the time Hahn left Montreal, in 1906, he had discovered new sources of radioactivity, notably forms of radium, thorium, and uranium, leading Rutherford to remark, "Hahn has a special nose for discovering new elements".[6]

Hahn returned to Germany, where he joined Emil Fischer's chemistry department at the Kaiser Wilhelm Institute in Berlin. He had turned down a career in industry and been "transmuted" from an organic chemist to one interested in the new field of radiochemistry, which straddles both chemistry and physics.[7] For radioactive measurements at the Institute, he used what had formerly been a carpenter's shop, while for regular chemical research he was allowed to work in the private lab of the departmental head, Professor Stock.

At that time, Hahn was interested primarily in thorium and its transformation properties. Although the most common isotope of thorium with a half-life of fourteen billion years appears to be almost stable, there are other isotopes that decay relatively quickly. Knoffer and Co.'s thorium factory provided what he wanted, including some almost pure thorium compounds that had been prepared at different dates. He found that the older the sample, the smaller its radioactive intensity, and he determined the amount of activity decayed to about half its initial value after a few years. However, one sample of substantially greater age showed a higher activity. He realised that this linked with his previous discovery of radiothorium. This is how.

Whenever the thorium compound was prepared, some radiothorium was present. Radiothorium has a half-life of about two years, so initially its activity diminished. Hahn deduced that as the radiothorium decayed, it was producing some hitherto unknown radioactive product with a longer half-life. Instead of the samples decaying uniformly over time, activity in some of the older ones had *increased* due to the accumulating presence of this new stuff. He used his chemical expertise to extract the new element, which he named mesothorium. He further found that he could separate mesothorium into two substances. One of these with a half-life of nearly seven years he called mesothorium 1; the other with a half-life of just six hours he dubbed mesothorium 2.

With radiothorium and now mesothorium—and in two varieties no less, mesothorium 1 and 2—Hahn was living up to Rutherford's observation about his ability to discover new elements. Chemists in Germany who had yet to fully appreciate Rutherford and Soddy's atomic disintegration hypothesis were less impressed. Emil Fischer himself found it hard to accept that radioactivity could detect substances in amounts far below what could be weighed. Fischer insisted that the most sensitive test for some substances was afforded by the sense of smell and that there was "no more delicate test than that!"[8]

Hahn's quest to find a "mother" substance that spawned radiothorium, and to identify the element that gives birth to radium by using quantities of matter so small they could only be detected by their radiations, was regarded by several Berlin chemists as little more than the work of a charlatan. The local physicists, who had reacted positively to Rutherford and Soddy's claims, realised the importance of the emerging field of radiochemistry and were more sympathetic to Hahn's obsession. As a result, Hahn began to spend time in the physics department attending classes on radioactivity and atomic transmutation.

He had spent the summer of 1907 trying to find the source of radium only to learn that Boltwood, the Yale radiochemist who had earlier described Hahn as the discoverer of a "compound of thorium and stupidity", had isolated it. Boltwood named the new "element" ionium—today we know this to be yet another isotope of thorium.

The summer over, classes in the physics department restarted, and Hahn went along. Among the handful of people in the audience that autumn he noticed a lone young woman, a rarity given strict rules which limited women's rights to enter the Institute's laboratories. When he enquired, he discovered her to be Lise Meitner, an Austrian who—inspired by the already iconic work of Marie Curie—had become interested in radium. So began a research collaboration spanning three decades between Hahn and Meitner, whom Albert Einstein would describe as "the German Marie Curie".[9]

LISE MEITNER

Meitner, then twenty-nine years old and four months older than Hahn, was dark, petite, and shy. Contrary to Einstein's description, she was Austrian, not German, and had just earned her PhD at the University of Vienna—only the second woman to do so. She specialised in the physics of radioactivity and in Vienna had already published two papers on the subject. Newly arrived in Berlin, and a lone woman in a man's world, she was desperate to continue her experiments. Hahn felt at ease in the company of women. He also admired Meitner's scientific enthusiasm and realised her expertise in physics would complement his in chemistry, making her the ideal research companion.

In those days women were not allowed to work in the Institute. Hahn lobbied the head of the physics department to allow her to join him. Finally, he won a concession: Lise Meitner could assist him in radium research in the carpenter's shop on the ground floor. Emil Fischer asked that she didn't enter the laboratories or study rooms on the upper floor where male students were working as that would set a precedent.[10]

Hahn had accumulated an almost complete collection of radioactive elements, including some strongly active actinium. Actinium had been discovered in 1899 by French chemist André-Louis Debierne, a colleague of the Curies. After they had extracted radium from pitchblende, Debierne analysed the residue and found the new element, a silvery metal, which glows in the dark with a pale blue light and is about one hundred times more radioactive than radium.

Actinium is now known to have atomic number 89, meaning that it occupies the eighty-ninth position in the periodic table of elements, three places below uranium at 92.[11] Back then, actinium was something of a mystery. The sequential decays of uranium to thorium, radium, radon, polonium, and lead gave some explanation for the appearance of these radioactive elements in uranium ores, though the production mechanisms were yet to be understood. Actinium, however, was an outlier. Its presence in pitchblende suggested it too was a descendant of uranium, though how was not yet known. A difficulty in interpreting these data in 1907 was that nothing was known about chemical elements being able to have multiple radioactive forms.

When Hahn and Meitner began their collaboration in 1907, the field of radioactive "elements" was growing, both in the number of examples and in the confusion generated when attempting to understand their taxonomy. During the next six years the pair spent long hours in the laboratory making precise measurements of half-lives. As atomic spectra are like a barcode for identifying the atomic elements, half-lives became the means for Hahn and Meitner to distinguish different radioactive forms of a single chemical element. This helped confirm the consensus of radiochemists that actinium precedes radium in the decay series—in other words, its decays lead eventually to radium rather than the other way around—but the mystery of actinium's parentage remained. As there was no sign of it being produced by thorium—radium's parent—Hahn suspected that there must be some yet undiscovered source, an element that he named protactinium. He and Meitner found some evidence for this yet unknown element, the gap between thorium and uranium in the periodic table that Mendeleev had dubbed eka-tantalum, but their work was interrupted by the start of the First World War.[12]

During the war, both Hahn's and Meitner's scientific knowledge was used by the German army, though to very different ends. Hahn was a lieutenant in the infantry reserve and his chemical expertise was infamously used in the deployment of chlorine gas. To his credit, he was so appalled when he witnessed the death agonies of poisoned Russians that he voluntarily became a human guinea pig testing the effectiveness of gas masks. Meitner, meanwhile, put her scientific skills to positive ends by using X-rays in medical work on behalf of injured troops.

ISOTOPES

Frederick Soddy's collaboration with Rutherford had convinced him of the power of radiochemistry, but he was still at heart a classic chemist. A special skill was his ability to analyse samples and determine their chemical composition, not least to forensically identify elements. He was convinced that radiothorium was chemically nothing more than thorium, not a novel element at all, and as for Hahn's two varieties of mesothorium, Soddy was confident these were decay products of thorium, one being chemically the same as radium, the other the element actinium. Yet from his expertise in radiochemistry, Soddy knew that each of these three manifested itself through alpha radioactivity in a different way to other atoms of the same element. For example, thorium having a half-life the age of the universe is normally barely radioactive, effectively appearing as a stable element in rocks, whereas the "radio" form has a half-life of but two years.

In 1913, Soddy's introduction of the concept of isotopes made the breakthrough that brought order to the chaos of radioactive "elements". What fundamentally distinguishes these isotopes was yet a mystery, but the concept was both correct and phenomenologically profound. Not least, it resolved a paradox that had emerged from other discoveries about atoms that had been made in the preceding three years. By 1913 it was clear that uranium is the heaviest naturally occurring element occupying position number 92 in the periodic table of elements with lead at 82, bismuth at 83, and several of the intervening elements such as actinium, thorium, radium, polonium, and radon already identified.[13] There were at most three gaps, at atomic numbers 91, 87, and 85, which was simply not enough room to accommodate the myriad of novel radioactive sources as distinct elements.[14]

Soddy had also confirmed by chemical analysis that alpha particles are electrically ionised examples of helium atoms. This complemented Rutherford's earlier work exciting a spectrum of helium, which had been physical. Now, the periodic table begins with hydrogen at position number 1, followed by helium at 2, so if uranium at number 92 emits an alpha particle—helium—two amounts of periodic displacement are lost to the uranium. This leaves the remnant atom as that at position number 90. And 90 is the location of thorium, which is indeed the empirical residue of uranium's alpha radioactivity. The link between alpha particles and helium brought order to the observation that alpha radioactivity transmuted an element into the neighbour that is two places before it in the periodic table.

This made sense too of other surprises that Rutherford and Soddy had found following their amazing evidence of one element transporting into another. Soddy had shown that he could separate out from thorium compounds a substance which appeared to contain much of the original thorium's radioactivity. They called it thorium X, although it was again chemically quite different from thorium (it proved to be chemically a form of radium). The sequence of events that Rutherford and Soddy had discovered now became clear: thorium first transmutes to this form of radium, which in turn changes into radon, at each stage spitting out alpha radiation. The last step in the chain of reasoning—that radiation is a direct result of transmutation—was possible thanks to Rutherford having shown in 1902 that alpha rays are particles of matter and Hahn's work with him demonstrating that these alpha particles are the same whatever their source.

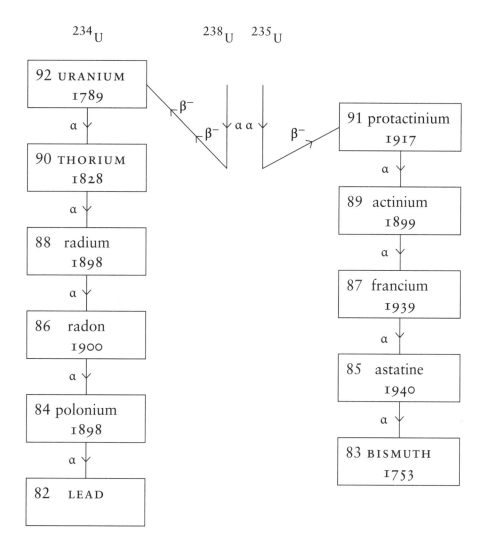

Figure 3. Radioactive ladders from uranium to stability. Uranium-238 and uranium-235 undergo alpha (α) and beta (β) decays via thorium-234 and thorium-231 to give uranium-234 and protactinium-231. (The postscript numbers refer to the total number of constituent protons and neutrons, or approximate masses in AMU, of these various isotopes.) A cascade of alpha decays leads to isotopes of lead and bismuth. Elements known before radioactivity are shown in capitals, and those discovered through radioactivity in lowercase. Year of discovery of most stable isotopes is shown.

Meanwhile in Berlin, by 1913 Hahn and Meitner had been studying beta radioactivity for five years. They established that the single charge of the beta (electron) was compensated by a change of one place in the element involved. So, beta radioactivity produces an element located one place earlier or later in the same periodic table. For example, the beta decay of thorium, element number 90, leads (after an alpha emission) to actinium, element number 89. Order was emerging, at least in the sense of categorising the heavy radioactive elements and filling in the highest rungs of the periodic table (Figure 3).

In addition to his expertise as a chemist, augmented by his work in radiochemistry, Soddy had come to his insight about isotopes following what is perhaps the most fundamental breakthrough of all: his former colleague Ernest Rutherford's discovery in 1910 of how atoms are built. In so doing Rutherford had located the source of radioactivity to be at the atom's heart in what today we call its nucleus. Radioactivity is thus truly *nuclear* energy.

4
The Nuclear Atom

To set the scene for Rutherford's epochal discovery of the atomic nucleus, rewind to 1906 when ten years had elapsed since Becquerel first detected the foggy smudge on his photographic plate that indicated the existence of radioactivity. During that decade scientific ingenuity had deduced this much about the strange phenomenon: First, atoms can radiate energy spontaneously in amounts far beyond anything explicable by chemistry; initially the behaviour had been identified as a property of uranium, but then a proliferation of radioactive sources had been found. Second, the laws of thermodynamics hold true: emission of energy involves change; atomic elements can transmute from one to another, facilitating that change. Hitherto unknown elements such as radium and polonium had been discovered, themselves radioactive descendants of uranium. Three distinct varieties of radioactivity had been identified, named alpha, beta, and gamma. Little was yet known about gamma radiation other than that gamma rays consist of very high energy particles of light. Alpha particles had been identified as ionised atoms of helium, beta particles as electrons.

Among the many unknowns three questions stood out: By what mechanism do atoms radiate energy in the form of alpha or beta particles? Where in the atom is this huge energy stored that is revealed solely—it would seem—by the phenomenon of radioactivity? And finally, as atoms can transmute from one element to another, atoms of different elements must have a common structure: what is it?

These were the "known-unknowns", but there would undoubtedly be many "unknown-unknowns" that answers to these questions would in time reveal. And whereas there were no clues on how to start attacking the first two of these known-unknowns, for the third—the nature of atomic structure—there was an opening. For this was the place where another discovery had provided the leverage that would reveal the atomic architecture and the bastions of radioactivity itself.

In 1897 J. J. Thomson had identified the electron as the negatively charged constituent of all atoms. This evidence of a new layer of reality had an immediate consequence. As gravity rules the cosmos, bulk matter made of atoms is electrically neutral overall, which implies there must be some positive charge inside atoms that neutralises the negative charge of their electrons. For the scientists of the day, the central question—where within atoms is the source of radioactive energy housed?—then inspired a more fundamental quest: what else besides electrons is inside an atom?

The year 1907 was when central characters in the drama were on the move, forming new collaborations that would in time change the direction of this research. As we have seen, Lise Meitner moved to Berlin and met Otto Hahn. Now Rutherford left Canada to become a professor at the University of Manchester where, with brilliant students and the great Danish theorist Niels Bohr, he would explain atomic structure.

He could not look directly inside an atom as it's far too small, so he had to deduce its makeup by implication. What he did is like trying to figure out the size and shape of an object hidden from view behind a screen, by firing pellets at it from one side and seeing how collisions with the target spray them around. For Rutherford, the pellets were alpha particles.

He didn't know how alpha particles are produced, but he didn't need to; it was sufficient to know that they moved fast, have a mass like that of a helium atom, and carry positive electric charge. For the yet unknown positive charges to entrap negatively charged electrons inside an atom, there must be powerful electric fields present, much as the sun's gravitational field holds the planets in their orbits. As alpha particles had been produced from within an atom, they must be small enough to reach an atom's innards where these electric fields are. Hence Rutherford's strategy was to first make a beam of alpha particles and a means to detect individual members, then direct the jet at atoms and see what happens.

To make the beam he placed a radioactive source of alpha particles inside a lead box. Lead absorbs alpha particles, but a small hole in one side of the box enabled a collimated beam to stream out. Now all he needed was a means of detecting the particles and the brains to interpret what he found.

SHELL SHOCK

The previous year, at McGill University, Rutherford had been studying the deflection of alpha particles by magnetic fields. He passed the alphas through a narrow slit, making a pencil-thin beam, which began to curve as the magnetic forces took effect. Downstream he placed a sheet of photographic paper to record when the alpha particles had hit. He covered the slit with a piece of mica—a mineral whose crystals can be split into extremely thin plates. In this case the sheet was a mere 1/300 centimetre thick. This was small enough to let the alpha particles through, but during their passage they nonetheless felt the effect of the atoms in the mica sheet. He noticed that the edges of the beam were now blurred. The change was like a targeted jet of water from a high-pressure hose being converted into a fine spray. The explanation in the case of the alpha particles was that the beam had been randomly scattered by atoms within the mica, sometimes considerably. This was astonishing because the particles were moving at about one twentieth the speed of light—if directed at the moon, uninterrupted they would reach it in just 20 seconds—with enormous energy for their size.

To get an idea of the significance of this, he used the most powerful magnets in his laboratory and let the beam traverse the magnetic field for about a metre. The deflection was little more than what the sliver of mica had already achieved. A metre compared to 1/300 centimetre: Rutherford calculated that within the mica there must be an electric field whose strength was about 10,000 volts per micron (millionths of a metre). Fields of such a strength in air would give sparks; the only explanation he could think of was that the powerful electric fields must exist with an exceedingly small region, smaller even than an atom.

This was the first hint that very powerful electric fields are present inside atoms. Rutherford speculated that these fields are what grip atomic electrons in place. To prove this experimentally is what he set out to do at Manchester.

To conduct the test would require him both to be able to count and to *see* individual alpha particles. To that end he worked with a German scientist on the faculty, Hans Geiger. They developed an electrical device that emitted a click every time an alpha particle arrived. This is what Geiger later developed into the modern Geiger counter. They were now able to count the alphas but also needed a means to locate them in space.

When alpha particles hit a screen coated in zinc sulphide, a faint flash of light is emitted in a phenomenon known as scintillation. The brief illumination is not a bright starburst like a particle boldly advertising its arrival but is instead barely perceptible. Rutherford had to cover the screen and his head with a linen sheet, like a classical photographer, to prevent stray light obscuring the dim ephemeral image. This was very stressful, so he shared the task with Geiger.

They directed the beam of alpha particles at pieces of aluminium, platinum, and gold, and painstakingly recorded the results. Rutherford pulled the cover over his head, watched intently, and every time an alpha particle hit, he recorded the fact in a foolscap logbook. Ten minutes of this left him exhausted, so Geiger took over to do his watch. Like a tag team they took measurements in turn, ten minutes on, ten minutes resting, the logbook showing maybe a dozen lines of data written in Geiger's fine script, interspersed with similar amounts in the rougher, bolder hand of the master.[1]

They discovered that the scintillations occurred at a single spot on the screen when the beam hit directly, but the spot became fuzzy after the beam had first traversed a thin sheet of mica—confirming what Rutherford had already seen back in Canada. His eye for the unexpected was troubled because in addition there was an incessant problem of stray scattered alphas, displaced from the general blur, that they could not explain. He wondered if these were a result of the alphas being deflected by the metal surface of the apparatus, but this seemed unlikely as it would require scattering through a large angle to get these strays into the region of interest of the main experiment. Nonetheless, he couldn't understand it. The strays were real and had to be checked.

Geiger was training a student named Ernest Marsden in the detection of radioactivity. He asked Rutherford if Marsden should begin a small research project. Rutherford agreed: "Why not let him see if any alpha particles can be scattered through a large angle?"[2]

To do so Marsden used an alpha particle beam directed at 45 degrees to a thin foil of gold. He placed one of Geiger's counters at 45 degrees to the foil so he could detect any alphas that bounced through a right angle. To ensure also that none had come straight from the source to the counter without hitting the target, he placed a lump of lead between the source and the counter. Alphas would not get through lead, so any that he recorded would have to have been scattered from the foil.

Gold leaf is about one hundred times thinner than human hair, so delicate that when held up to the light you can see through it. The idea that it would interrupt alpha particles travelling at 18,000 kilometres a second seemed fanciful. So Marsden was immediately surprised to find alphas being scattered into his detector. This showed that fast-moving alphas packing a considerable punch could nonetheless be turned through a right angle. He double-checked, triple-checked, until he was sure he had overlooked nothing. He then tried an even more extreme test by placing a lead screen adjacent and parallel to the alpha particle beam. If he were to detect any alphas now, they would have effectively been turned back almost in their tracks having hit the target. And to his astonishment he did, discovering that about one in twenty thousand alpha particles bounced right back from whence they had come. This was so incredible that Rutherford exclaimed: "It was as though you had fired a 15-inch shell at a piece of tissue paper, and it had bounced back and hit you".[3]

THE NUCLEUS REVEALED

Rutherford knew from the way that alpha particles move in a magnetic field that they are positively charged and had the characteristics of helium ions. The fundamental law of electric charges—like charges repel—implied that the positive charges within the target's atoms must repel the positively charged alpha particles. His interpretation of what happened was that occasionally this repulsive force had first slowed an alpha particle to a halt and then thrust it back from whence it came.

In his mind an image of the positive charge being concentrated in a massive lump at the centre of an atom began to form. Whereas the lightweight electrons hardly disturbed the alphas, he visualised the atom's positive charges being concentrated on a heavy ball at a point in the middle. For a heavy element such as gold this *nucleus* is about fifty times heavier than an alpha particle, so the atomic nucleus will remain unmoved while the alpha recoils, like a rubber ball bouncing back from a rock. The nearer the alpha particle is to making a direct hit, the more it will be deflected. In a head-on collision, it stops momentarily and then comes right back.

Remarkably, this sequence of events must have happened within less than a picosecond—a trillionth of a second—for that is how long speeding alpha particles would take to pass freely through gold leaf. Rutherford estimated that the electric force capable of giving an alpha particle a U-turn in that instant must be a thousand times stronger than anything known—far beyond what Geiger and Marsden had experienced in their experiments with mica.

Rutherford calculated just how close the alphas could get to the lump, and discovered that on rare occasions they could reach within a ten-thousandth of an atom's radius before being turned back. By measuring how often the alpha particles were deflected violently, were deflected moderately, or passed through the mica almost untouched, Rutherford deduced that the compact, positively charged nucleus contains almost all of an atom's mass but occupies less than a thousandth of a millionth of a millionth of its volume.

The negative electrons circulate around the nucleus at relatively large distances. As the nucleus is so small, most of the alphas go straight through or nearly so. Only rarely do they bounce out at a large angle. It was these occasional near hits that led to the large deflections Marsden had measured, and which had been revealed initially as the stray scatters in Rutherford and Geiger's first experiment.

Rutherford came to this conclusion in December 1910. Hans Geiger recalled him announcing it "just before Christmas" at a supper party in the Rutherford household. Manchester physicist Charles Darwin—grandson of the famous botanist—was present and recorded this to have been "one Sunday evening". Further clues to the date are in a letter Rutherford wrote on 14 December to Bertram Boltwood, where he remarked "I think I can derive an atom which explains Geiger and Marsden's results".[4] Put these together and we come to 11 December as the probable Sunday, if he told his colleagues before writing to Boltwood, or 18 December otherwise.

In any event, although Rutherford only went public in a talk to the Manchester Literary and Philosophical Society on 7 March 1911, the nuclear atom was discovered in 1910. In his formal lecture Rutherford described "a central electric charge concentrated at a point and surrounded by a uniform spherical distribution of opposite electricity equal in amount".[5]

So, by 1910 Rutherford had established the existence of a positively charged, compact, heavy atomic nucleus but remained ignorant of what it consisted of or how it was formed. And although radioactivity in the form of alpha particles had given Rutherford the tools with which to discover the atomic nucleus, he was no nearer to understanding radioactivity's cause, let alone that the atomic nucleus was key to the immense energies that radioactivity implied.

As for the surrounding spherical distribution of "opposite electricity", this consisted of Thomson's electrons, which left a new question: why are electrons not captured by the nucleus towards which they are attracted? Rutherford did not address this.

THE GREAT DANE

In 1911 Niels Bohr was a twenty-seven-year-old Danish theoretical physicist working at Cambridge University with J. J. Thomson. In December that year Rutherford visited to speak at the annual Cavendish dinner. Bohr was present and talked with him afterwards. He was galvanised by Rutherford's enthusiasm and willingness to listen to the ideas of a "young man". Soon after, Bohr visited Manchester and Rutherford was impressed. This was unusual because Rutherford famously disregarded theorists, but when asked later why he had warmed so quickly to Bohr, he said "Bohr's different. He's a football player!"[6] Indeed, Bohr had played in goal for one of Denmark's leading teams, and his brother Harald even played for the national team. The outcome was that in March 1912 Bohr moved to Manchester to learn about radioactivity and the nuclear atom firsthand.

Bohr was powerfully built, with a large dome-shaped head, and a soulful demeanour due to his substantial lips, which drooped at the edges. He spoke softly in a heavily accented monotone, which could be hard to hear. Even when audible, he was difficult to follow as he would think aloud, refining ideas in an elliptical conversation with himself rather than the listener, making a verbal sculpture until the polished explanation arrived. By such mental exertions, within a year Bohr had come up with a remarkable series of insights about what Rutherford's results implied for the nature of the nuclear atom. He had no forensic proof for any of these, but instead had forged a chain of circumstantial evidence that collectively was greater than its individual pieces—and, we now know, is basically correct.

First, radioactivity originates in the central nucleus whereas chemical properties depend upon the number and configuration of the electrons in the atom's periphery. Second, since the positive charge of the nucleus determines the number of negative electrons, and since the latter determines the chemistry, the nuclear charge must be the same as the position of the atomic element in Mendeleev's table, which was founded in the elements' chemical periodicity. Thus, a hydrogen atom must have a nuclear charge of one and contain one electron, helium with charge two contains two, and onwards up to uranium at ninety-two.

Meanwhile, also at Manchester, a twenty-six-year-old experimental physicist named Henry Moseley was measuring the energies of X-rays emitted when a beam of electrons scatters from metals. A beam of X-rays—high energy light—can be split by a crystal much like visible white light is split into colours by a prism. By this means the frequencies of the X-rays can be measured. Quantum theory implies that X-rays consist of particles—photons—with energy proportional to their frequency, whereby the energies that X-rays have taken from the metals' atoms are determined.

Moseley found that the most energetic X-rays followed a simple rule. Relative to the energy emitted in the case of hydrogen, the magnitudes in all cases turned out to be integers which increased in parallel to the atomic weight of the atomic element emitting the rays. The magnitudes of the atomic weights varied haphazardly, but Moseley had now numbered the atomic elements. Inspired by Bohr's model, Moseley concluded that these integers—the element's atomic number—must be the number of charges on the atomic nucleus of the relevant element.

Moseley would have won a Nobel Prize for this breakthrough had he not been killed at Gallipoli in 1915 having joined the army at the start of the First World War. By this stage, 1913, thanks to Rutherford, Bohr, and Moseley having unravelled atomic structure, there was clear proof that uranium, thorium, and lead have atomic numbers 92, 90, and 82, respectively. This determined their positions in the periodic table and quantified the number of elements that remained to be assigned.

This was almost exactly the time when Hahn, Meitner, and the community of radiochemists were troubled by finding far more radioactive elements than there were spaces in the periodic table. This is where Moseley and Bohr's focus on atomic number, and Soddy's hypothesis of isotopes, instantly brought order to the nuclear jungle. Soddy's new insight that atoms with the same chemical properties could have different radioactive behaviour was taken up by Bohr who explained the sequences of radioactive transmutations that the radiochemists had uncovered. Isotopes of an element all have the same atomic number, and therefore, he announced, when a radioactive decay occurs by emitting an alpha particle, a helium ion with two positive charges, the element shifts two places backwards in the table. If it emits a beta particle, with a single negative charge, the element moves one place forwards. If it absorbs a negatively charged beta particle (or, as later realised, emits a positively charged positron), the move is one place backwards—recall Figure 3 in Chapter 3.

Bohr had thus produced the first rational picture linking Rutherford's nuclear atom to the places of elements in the periodic table and had also included isotopes in that phenomenology. As to why there are isotopes, add that to the list of unknowns. One thing was clear, however: atomic nuclei are distinguished by more than just electrical charge. Like Rutherford, Bohr had no insight into the inner labyrinths of that paradoxically compact nucleus, on the one hand infinitesimally small and yet on the other carrying the bulk of an atom's mass together with an immense concentration of positive electric charge. How the electrons encircled the central nucleus, however, was a question that Bohr now addressed.

THE ATOMIC BAR CODE

Bohr's vision was of a massive static atomic nucleus surrounded by several flighty lightweight electrons. In popular imagination, Bohr's atom is often pictured as a miniature solar system. In this naive analogy, the nucleus plays the role of the sun, and electrons are like the remote planets. Whereas the force of gravity controls the motion of the latter, it is the electrical attraction of opposite charges—positively charged nucleus and negatively charged electrons—that holds atoms together.

Analogies can be dangerous if stretched too far, and the case of the planetary electrons is a cautionary example: atoms built like that could not survive for a moment if they obeyed Isaac Newton's laws of mechanics. Had electrons in atoms encircled the nucleus like planets orbiting the sun, and obeyed Newton's laws, they would have lost energy by radiating light and spiralled into the nucleus within a mere fraction of a second. An atom, once formed, would self-destruct in a flash of light almost immediately; matter, including you and I, would not exist.

Bohr was aware of this problem, and his genius was to find a solution. The fact that we are here shows that very small things, such as atoms and their constituent parts, follow different laws from those of Newton, which explain the behaviour of objects that are large enough to see. Today we know these laws: on the atomic scale, Newton's classical mechanics give way to *quantum* mechanics. In quantum theory, an atomic electron cannot go wherever it pleases; instead it is limited, like someone on a ladder who can only step on individual rungs. Electrons in atoms follow a fundamental regularity, each rung corresponding to a state where the electron has a unique amount of energy. Bohr had this epiphany in the summer of 1912, his inspiration being a remarkable observation, in 1885, by a Swiss schoolteacher, Johann Balmer.

Light, that rainbow or spectrum of colours, consists of electromagnetic waves whose electric and magnetic fields oscillate hundreds of trillions of times each second; what we perceive as colour is the brain's response to the different frequencies of these oscillations. Albert Einstein—most famous for his theory of relativity—won his Nobel Prize in 1921 for showing that light rays, rather than being a continuous stream, consist of a staccato burst of particles, called photons. A photon has no mass, but travelling at the speed of light it has energy. A photon at the high-frequency violet end of the rainbow has roughly twice the energy of one from the low-frequency red end.

Lamps shine because heat—of a tungsten filament, or of neon or mercury vapour—is shaking photons loose from the source's atoms; they have characteristic colours because the photons emerge with energies, or frequencies, unique to the parent atoms of each element. These colours identify the pattern of energy levels available to the electrons within those atoms. When an electron drops from a rung with high energy to one that is lower down, the excess energy is carried away by a photon of light. By viewing light through a diffraction grating—a piece of glass that has been scratched with many closely packed grooves—it is possible to split it into a spectrum of component colours, such that sharp bright lines become visible.

The lines are like some fundamental barcode identifying the elements generating that light. The question is: how do we crack the code?

In 1885, Balmer discovered a remarkable feature about hydrogen's spectrum: the frequencies of its lines fit a simple formula, each being proportional to a common quantity multiplied by the difference of two numbers, which themselves followed a simple rule. These two numbers were $1/4$—written as $(1/2)^2$—and $1/n^2$—where $n = 3, 4, 5$, etc. Balmer's simple formula described the spectrum of the hydrogen atom perfectly, but not its cause. By luck or judgment, Balmer had stumbled upon a great truth. The question was: why does his magical rule work?

In 1912 Bohr found the explanation, courtesy of the new quantum theory. In quantum theory, an atomic electron's orbital angular momentum is restricted to discrete amounts. These values are integer multiples, n, of a basic quantity (quantum), proportional to Planck's constant. Bohr calculated that the energies of these constrained orbitals are proportional to $1/n^2$. The energies of the rungs on Bohr's ladder miraculously explained Balmer's discovery; when an electron drops from a high-energy rung to a lower one, the difference in energies is radiated as light in accord with Balmer's formula.

Later, in 1924, the French aristocrat Prince Louis de Broglie gave a dynamic picture of this in his PhD thesis at the Sorbonne. He proposed that in quantum theory any particle can take on a wavelike character; what is familiar for photons and electromagnetic waves occurs for electrons also. Visualise the waves for electrons in atoms as if they were wobbles on a length of rope. When coiled in a circle, like a lasso, for a wave to fit perfectly into its circumference, the number of wavelengths in the circuit must be an integer. Imagine this circle like a clockface. If the wave peaks at twelve o'clock, with a dip at six o'clock, the next peak will occur perfectly at twelve: the wave "fits" into the circle. However, a peak at twelve followed by a dip at five o'clock would have its next peak at ten and be out of time with the beat of the wave—*out of phase* in the jargon of physics: the wave will not "fit". Electrons circulating in atoms cannot go anywhere they please but only on those paths where their waves fit perfectly on the lasso. The numbers in Balmer's formula turn out to be related to the numbers of wavelengths in a single circuit.

Bohr had discovered the basic rules governing electrons, which underpin chemistry. The implication was so profound that in 1922 this won him the Nobel Prize for physics. A common quiz question is whether Rutherford won the physics prize for his discovery of the atomic nucleus. The answer, perhaps surprisingly, is no. Rutherford had won the 1908 Nobel Prize for his work on *radioactivity*, and it was for *chemistry*. Like Einstein and relativity, his most famous and far-reaching discovery won no Nobel recognition.

Rutherford had identified the nuclear atom from experiment, while Bohr's contemplation had put the pieces together and theoretically explained the periodic table, radioactive series, and more. The atomic nucleus had been revealed through the chance discovery of radioactivity followed by experimental and theoretical genius. How a compact lump of positive electricity could be held so tightly, storing energy that could leak in radioactivity for aeons and, moreover, form isotopes, stayed a mystery. The nature of the enigmatic nucleus remained to be explained.

PART II

The Nucleus Explained
1914–1932

5
Rutherford "Splits the Atom"

"If, as I have reason to believe, I have disintegrated the nucleus of the atom, this is of greater significance than the war."[1]

The secretary of the British government's committee on antisubmarine warfare read the apology for absence with astonishment. It was 1917, and world war had been raging for three years. In addition to the horrors of the trenches, there was a war at sea, where German U-boats were menacing the British Navy and merchant vessels. Two years earlier the writer of the apology, the newly knighted Sir Ernest Rutherford, had produced a top-secret report on the possibility of using acoustic technology, a signalling system that would use sound waves beyond the range of human hearing, to detect these submarines. This was the first mention of the system that would one day become modern sonar. Its existence would remain an official secret for twenty years.

Rutherford and colleagues on the committee had performed classified experiments to test underwater microphones known as hydrophones. Following Rutherford's 1915 report, the first tests were made in water tanks in laboratories at the University of Manchester, where he was based. Recently his team had been using fishing trawlers to conduct full-scale tests at a research outpost in Fife on the east coast of Scotland. They had produced a working prototype of sonar, which the British originally called ASDIC—short for "Active Sound Detection supersonICs". Yet although British Navy warships were being outfitted with early sonar prototypes, the technology was not heavily used until the Second World War.

While Rutherford's main priority was to apply his scientific skills to the war effort, he did manage to work on atomic physics sporadically. By 1914, he had established the presence of the atomic nucleus but had no idea of its makeup, for although his alpha particles were travelling at one twentieth the speed of light, they were too feeble to penetrate *within* the nucleus. To have any chance of invading the nucleus itself, Rutherford had three courses of action: either he must speed up the alpha particles to prevent them being so easily kicked aside by the nuclear electric field, develop more intense sources to increase the chance of some getting through, or irradiate lighter elements for which the resistance is less. Lacking any practical means of accelerating the alpha particles to explore the nuclei of heavy elements, nor the expertise for creating brighter sources, Rutherford turned his attention to the third option.

H-PARTICLES

Bohr's identification of atomic number with the number of positive charges in the nucleus, hydrogen the simplest having one positive unit of charge, gave a vision of the hydrogen nucleus being the prototypical positive charge common to all nuclei. The atomic number of an element then signifies the number of these *H-particles* in its nucleus. As like charges repel, however, the natural tendency would be for electrical repulsion to make such a nucleus self-destruct. To bind it requires some strong attractive force capable of generating huge amounts of energy. There was no idea how this comes about.

Rutherford, meanwhile, was certain that the energy revealed through radioactivity "is in the atom itself", and Bohr's interpretation of Rutherford's nuclear atom implied this to be nuclear energy.[2] From this it was a small step to conclude that radioactivity is a manifestation of the energy responsible for holding nuclei together. As water runs downhill to gain stability by lowering its potential energy in Earth's gravitational field, so can an isotope which requires lots of nuclear energy to hold its positive charges in place, convert to another more stable formation which needs less energy. The difference in the two "binding energies" is what is manifested as the energy in radioactivity.

These theoretical images were compelling, but initially that was all they were—theoretical. How to test them experimentally and reveal the nuclear framework was the question. It would be an experiment with Marsden in Manchester that established the isolated hydrogen nucleus as the fundamental unit of positive electrical charge in atomic nuclei and proved to be the first breakthrough in establishing their internal structure.

The discovery had come about as a result of Marsden firing alpha particles at hydrogen gas in 1913. Rutherford suggested that he do this, expecting the results to be very different from what Marsden had earlier found when alpha particles hit the atoms of heavy elements. Alpha particles are much lighter than the nuclei of gold or of other atomic elements they had studied but are four times the putative mass of the yet hypothetical H-particles. The difference between the two cases is like the contrast between a lightweight rubber ball hitting a more massive football and its converse, a football smashing into a small stationary ball. In the first case, the football is barely disturbed whereas the rubber ball bounces off violently, even returning along its original path on some occasions if there is a direct hit; this is the analogue of Marsden's original observation of large-angle scattering of alpha particles from heavy nuclei. In the second case, the football maintains much of its own motion while the previously stationary small ball is projected rapidly forwards along the direction of the incident football. This is what the impact of an alpha particle on an H-particle would be like: the fourfold heft of the alpha particle would carry it onwards, but slower, while the relatively light H-particle would be projected quickly forwards.

Rutherford knew that successive collisions with atoms of hydrogen gas would slow the alpha particles until they come to rest. As the alpha particles come from a common radioactive source, they all set off with the same energy and so come to a halt at about the same distance—known as their range. To detect these particles, Rutherford and Marsden used a zinc sulphide screen, as before, and varied its distance from the source. Beyond their range, alpha particles can no longer penetrate the gas to strike the screen; scintillations should cease.

But the two scientists found a few scintillations *did* occur beyond the range of the alphas, and with a more point-like character than the flashes that signalled alpha particles. Whatever was responsible could travel up to four times further through the gas, which is consistent with them having one quarter the mass of an alpha particle. Scintillations are caused by charged particles, and deflection by a magnetic field showed this charge to be positive. Rutherford concluded that these could only be the nuclei of hydrogen, knocked out from atoms in the gas by the energetic alphas.

It took a few more years before he established that the H-particle is a fundamental building block occurring in the nuclei of *all* atoms. This came about once again by Rutherford worrying about a small detail, an anomaly that Marsden noticed in their otherwise successful experiment.

In summer 1914 Marsden began to have trouble with his experiments. He noticed that when the scintillation screen was outside the tube such that the H-particles travelled through the air, more of them appeared than the impact of alpha particles on the hydrogen alone could account for. Were these H-particles being produced not just from hydrogen but from the air too? Unfortunately, Marsden left Manchester that autumn to take up a teaching post in New Zealand, and then world war began causing scientific research for the war effort to take over much of Rutherford's time. Nonetheless, he spent any spare moments on pursuing the anomaly.

"A TREMENDOUS DISCOVERY"

Following Marsden's departure, Rutherford investigated the puzzle himself, whenever he found half a day to do some research of his own. The experiment typified Rutherford's style, where he used simple concepts to tease out profound truths. His apparatus consisted of a small brass box, with two stopcocks allowing gas to be evacuated or admitted. In the middle of the box was a small disc of radium salt, which was the source of alpha particles. At one end was a scintillation screen made of zinc sulphide beyond which was a microscope enabling Rutherford to study the faint flashes carefully.

Initially he repeated what Marsden had done, checking that alpha bombardment of hydrogen produced scintillations at the same range and of the same character as Marsden had seen. His immediate goal was to eliminate the possibility that any H-particles being produced were the result of small traces of hydrogen that are present in air normally.

He evacuated all air from the box and input dry oxygen. It had to be dry so that there was no water vapour—H_2O—giving rise to unwanted hydrogen. He found that the alpha particles were absorbed, their range being as expected, and there was no sign of H-particles. He repeated the experiment with dry carbon dioxide and the results were the same: no H-particles. Then he tried using dry air. In this case he saw clear scintillations with the character of H-particles and moreover at a range typical of them; Marsden's anomaly, where air seemed to generate H-particles, had returned. Worried that there might nonetheless be some water vapour in the atmosphere he dried the air more by passing it through plugs of absorbent cotton, but the results were the same as before: a clear sign of H-particles.

What does air contain beyond oxygen and carbon dioxide? Rutherford knew the answer: air is 78 per cent nitrogen. So, he decided to repeat the experiment, this time with pure nitrogen gas, and compare the scintillations with those that he had found with air. His intuition was right. The number of H-particle scintillations when he used pure nitrogen gas increased dramatically.

By now it was 1917. He already had a fair idea of what his results implied, though it wouldn't be until after the war's end that he had time to do more tests and be absolutely certain. His preliminary opinion was that "so far" the results suggested that when alpha particles collide with nitrogen, "the resulting long-range atoms are not nitrogen atoms but probably atoms of hydrogen". And then the shattering implication, literally: "If this be the case, we must conclude that the nitrogen atom is disintegrated."[3] Testament to his confidence was a letter he wrote to Bohr on 17 December saying he was "trying to break the atom", meanwhile urging Bohr to keep this news private.[4]

Rutherford had reached the threshold of resolving what an atomic nucleus consists of. Building on Bohr's hypothesis, he concluded that a hydrogen atom has the simplest nucleus of all, consisting of a single positively charged particle—the H-particle—which he now named *proton*. He postulated that the proton is fundamental to the nuclei of all atomic elements, that the number of protons in a nucleus determines its total electric charge and thereby its position in the periodic table of elements. For example, helium, element number 2, contains two protons; nitrogen, number 7, contains seven; and so on.

This is what Rutherford interpreted had happened when he used alpha particles to bombard nitrogen. If the nucleus of each nitrogen atom indeed contains seven protons, and if one is hit by an alpha particle containing two, that gives nine in all. If the impact ejects one proton from the nitrogen and the alpha particle attaches to what remains, there will be a lump containing eight protons, the nucleus of the eighth element in the periodic table: oxygen. Here for the first time was an artificial transmutation of atoms (Figure 4).

By detecting the ejected proton, Rutherford had identified a key component of the atomic nucleus and the source of the electric field

Protons	2 + 7	→	1 + 8
Mass	4 + 14	→	1 + 17
Nuclei	α + N	→	p + O

Figure 4. Transmutation of nitrogen into oxygen. This "splitting" is more accurately described as a building-up of atoms as oxygen with eight protons and atomic mass 17 AMU has been produced from nitrogen's seven protons and mass 14 AMU.

that envelops the atom. Rutherford had already won a Nobel Prize for his work that helped establish the existence of huge reserves of energy deep within the atomic nucleus. Now he had for the first time disintegrated the nucleus of an atom and set in motion the processes which would lead to the explosive release of that energy.

He had for more than a decade been aware of this vast energy store, and rightly foresaw that nuclear disintegration would be of greater significance than the immediate needs of the allies in the World War. He had expressed his hopes and fears about nuclear energy in a public talk in February 1916 where he warned "at the present time we have not found a method of dealing with these forces and personally I am very hopeful we should not discover it until man is living at peace with his neighbours".[5]

Rutherford had been thinking about nuclear structure for so long that it seems he was less surprised at the fragmentation of the nitrogen atom than at the robustness of the alpha particle. He remarked: "Considering the enormous intensity of the forces brought into play, it is not so much a matter of surprise that the nitrogen atom should suffer disintegration as that the alpha particle itself escapes disruption into its constituents". And, he added, "if alpha particles—or similar projectiles—of still greater energy were available for experiment, we might expect to break down the nucleus structure of many lighter atoms".[6]

He finally completed his forensic examinations and published his results in April 1919, shortly before he moved to Cambridge to succeed J. J. Thomson as head of the Cavendish Laboratory. Initially this remarkable discovery made little public notice outside the scientific community. The change came about when Charles Nordmann, a French astronomer and populariser of science, visited Manchester in 1919. The result of his meeting with Rutherford was a front-page article for the Paris newspaper *Le Matin* on 8 December 1919 headlined "*Une immense découverte*" ("A tremendous discovery") whose "consequences, theoretical and practical, are incalculable". The article, which included a picture of a moustached Rutherford, sporting a wing collar, carried the byline "*La pierre philosophale est trouvée*" ("The philosopher's stone has been found").

Nordmann explained how the difference between an atom of iron and one of gold is just the number of electrons and of the positive particles in the nucleus. He added that previously the spontaneous transmutation of elements in radioactivity had been determined by nature, over which humans have no control, and scientists were still unable to speed or slow the rate of these atomic transformations. But now Rutherford had successfully dissected a stable atom. By bombarding nitrogen gas with alpha particles travelling at 20,000 kilometres a second, he could convert some nitrogen into hydrogen. This was the first transmutation induced by humans, and in stable matter no less.

Rutherford's success confirmed also that those atomic elements are built from common constituents. Nordmann now conjectured that by bombarding lead with alpha particles it could be possible to convert some of its atoms into gold. The cost of doing so, however, would likely exceed the value of the gold that you obtained. Nordmann's scoop was picked up by the Press Association, and within days the worldwide English-language media were trumpeting the news: "Rutherford splits the atom".

A NEUTRAL ATOM

The Bakerian Medal and Lecture is the Royal Society of London's premier lecture in the physical sciences. Funded by Henry Baker FRS in 1775, it has been given annually ever since. In 1920 Ernest Rutherford was the scientist honoured by unanimous choice to make the presentation.

Nordmann's description of Rutherford's disintegration of the atomic nucleus as an immense discovery was endorsed by the scientific community; his hypothesis that the proton is the fundamental source of the positive charge of all atomic elements had further revolutionised atomic physics. Scientists thought that in the massive proton and the flighty electron together with the electrically neutral photon, the massless corpuscle of light, they had at last identified the basic seeds of matter. The Bakerian Lecture would be the ideal opportunity to have Rutherford explain what this implied and to hear him present his vision for the future of the field.

Having chipped the proton out of the nucleus of the nitrogen atom, Rutherford was already trying similar tests with other light elements. Thanks to this simple experiment, Rutherford had explained half of nuclear physics—"half" because although the hypothesis that the number of protons determines the order of the elements in the periodic table was robust, protons alone could not match the relative masses of the elemental atoms (recall Figure 4).

Understanding both the electric charge and masses of atomic nuclei required not just the proton but also some ingredient that contributes mass without electric charge: helium, for example, with a net charge of two clearly required *two* positives, but its atomic mass is some *four* times that of hydrogen. Rutherford was immediately caught in a logical trap, thinking that the nucleus contained both positive and negative charges. His hypothesis was that a helium nucleus—an alpha particle—consists of four massive positive protons and two lightweight, negative electrons.

Rutherford's belief that there are electrons within the nucleus seemed logical because electrons are ejected in beta radioactivity, which is a nuclear event, and how else could electrons come from a nucleus other than by being there already? The answer is that they can be created by forces within the system: like how a dog's bark is created using air in its lungs, for example. In 1920, however, this was an understandable error, and the solution of how electrons emerge in radioactivity would not be found for another decade. Now, in the Bakerian Lecture on 3 June 1920, Rutherford developed this by proposing that an electron and proton could form a compact neutral entity with a mass like that of the proton, making "an atom of mass one which has zero nucleus charge". Rutherford added: "Such an atomic structure seems by no means impossible." Today we refer to this as the *neutron*—the electrically neutral counterpart to the proton.[7]

Rutherford pointed out that at rest the neutron would be a natural sibling to the proton in atomic nuclei, adding mass to an atom without changing its chemical properties. He predicted isotopes of hydrogen where its single proton is joined with one or two neutrons. These nuclei of "heavy hydrogen" are today known as the deuteron and triton, the nuclear seeds of deuterium and tritium.[8]

On the move, such a neutral particle would have a remarkable ability to penetrate nuclei since its motion would not be impeded by the electrical charge of the nucleus. His own words are revealing: "Such a [neutral] atom would have very novel properties. Its external field would be practically zero except very close to the nucleus, and in consequence it should be able to move freely through matter. Its presence would probably be difficult to detect, on the other hand it should enter readily the structure of atoms and may either unite with the nucleus or be disintegrated by its intense field."[9]

He added that the existence of such atoms "seems almost necessary to explain the building up of the nuclei of heavy elements": the property of adding mass without charge is the source of isotopes. To visualise what he has done, imagine for a moment that a proton is like a single red snooker ball and that a bag of 92 red balls, representing the protons that make a uranium nucleus, mysteriously turns out to weigh 240 times that of a single ball. Rutherford in effect deduced there must be something else in the bag—white snooker balls—each of which weighs the same as a red one. Put 148 white ones together with the 92 red ones, and this will explain the 240-fold heft of the total. The white balls are, of course, neutrons.

His insights that the neutron should "enter readily the structure of atoms", as there would be no electrical barrier, unlike the case of alpha particles where positive charge is repelled by the positive charge of the nucleus, would later be key to liberating nuclear energy explosively. So far, so good, but mentally caged by the concept that nature makes do with just the electrically charged proton and electron, and a neutral photon, Rutherford supposed his massive neutron to be a close-knit pairing of a positive proton and a negative electron. This would be one of Rutherford's rare failures of intuition. The neutron exists, we now know, but not as some mingling of a proton and electron. Instead, it is an electrically neutral particle no less fundamental than its near-twin, the proton.

Rutherford's wrong picture led him to suggest that upon meeting a heavy nucleus the neutron might "be disintegrated [into a proton and electron] by [the nucleus's] intense field". He correctly predicted that the particle would be difficult to detect, but in suggesting it could be disintegrated into these components, he inadvertently set physicists on a wild goose chase for over ten years. The person most affected by this error was James Chadwick.

As a student at Manchester University, in 1912, Chadwick had heard Rutherford's discourse on the nucleus. He had started research in nuclear physics but was working in Berlin when the First World War began, which led to him being interned for four years. After the war he returned to England to continue working under Rutherford's tutelage. In 1919, when Rutherford moved to Cambridge, he encouraged Chadwick to join him with an oft-repeated clarion call: "I have done enough for one man; it is now the task of the younger generation to tackle the problems about the structure of the nucleus".[10] Chadwick duly moved to the Cavendish Laboratory, where he earned his PhD and at the age of thirty became Rutherford's assistant director of research. He was in pole position to pursue the neutron and began to search for it in 1921, inspired by Rutherford's compelling arguments for the particle's existence. From the start, however, Chadwick designed his experiments based on Rutherford's erroneous belief that his quarry was a composite of a proton and an electron.

In a hydrogen atom, the electron is very far from the proton. An idea of how far is if we were to imagine that atom's breadth scaled up to the length of the longest fairway on a major golf course. If the proton is envisioned to be like the hole, some 5 centimetres across, the electron would be some 500 metres away, probably more remote even than the tee from where the golfer makes their initial drive. Chadwick had tried blasting atoms with powerful electric discharges or exposing them at high altitudes to bombardment by cosmic rays, in the hope of kicking an electron into the proton in a veritable hole in one. He patiently looked for gamma radiation produced in the moment of their combination but saw none. Had he succeeded he would have brought together what nature had cast asunder, for hydrogen atoms have existed like that forever, the electrical attraction of electron and proton being perfectly balanced when so separated. As there was no known force capable of gripping the pair together, the attempts to fuse electrons and protons sound like those of a madcap scientist trying to undo billions of years of natural construction. Chadwick would later describe his attempts to synthesise a neutron from this pair as "quite wildly absurd".[11]

The constraints of quantum physics dictate how electrons and protons settle naturally at an atom's length apart. This theory was not well understood then, certainly not by Rutherford who preferred experimental observations to lead and theorists to follow. It is probably because he had no idea as to how an electron and proton might fuse together that he never formally published his proposal in a research paper, instead leaving it a remark made in his *tour d'horizon* Bakerian Lecture of 1920. In any event, Chadwick failed and by 1928 had given up.

6
The Mystery of Beryllium

MENDELEEV'S MISSING ELEMENT

For two years during the First World War, while working as an X-ray technician supporting German troops on the Eastern Front in Poland, Lise Meitner arranged her leaves of absence to coincide with those of Otto Hahn and rushed off to Berlin's Kaiser Wilhelm Institute to work with him. In 1917 she returned to Berlin. With most of the men having joined the army, she was appointed head of her own physics laboratory. Hahn visited her during a period of leave that year, and they reviewed their work on radioactivity.

Back in 1914, before the war intervened, they had been trying to find the mother substance of actinium. With Meitner now able to work uninterrupted in Berlin, they agreed that she should continue the quest. Hahn's memoir described their goal: "To find the substance which forms the starting point for the actinium series [of sequential alpha decays from actinium producing the stable element bismuth] and to determine through which intermediates actinium is derived".[1]

There were clues that the source was the element predicted by Mendeleev's table to occur between uranium, atomic number 92, and thorium, at 90. First, in 1900 William Crookes had found an intensely radioactive material coming from decays of uranium, but he had not identified it. Then, in 1913 the Polish chemist Kasimir Fajans and his German student Oswald Göhring detected something with a half-life of a mere 70 seconds produced in beta decay of uranium; they named it brevium, but it didn't live long enough for them to do chemical tests and identify it with Mendeleev's prediction. By 1917, Bohr's theory that beta and alpha radioactivity change atomic numbers by respectively one or two had become lore, so beta decay of uranium at number 92 would produce element number 91. This further convinced Hahn and Meitner that in beta decays of uranium, the missing element was there to be isolated and confirmed.

They also had some indirect evidence of their own, as in 1914 they had measured the alpha particle spectra from pitchblende and found one example that came from no known substance. Chemical analysis of the ore also revealed the presence of actinium. Everything was theoretically consistent with the pitchblende's uranium—atomic number 92—having produced what would become known as protactinium—atomic number 91—which then decayed by alpha emission moving two places down the periodic table to leave the actinium—number 89. All fitted with the hypothesis that uranium had produced the protactinium by beta decay, the protactinium in turn emitting the alpha particle. This was reminiscent of Sherlock Holmes's advice to Dr Watson that "when you have eliminated all which is impossible, then whatever remains, however improbable, must be the truth".[2] What Hahn and Meitner had found was indeed encouraging, maybe even convincing, but it was not yet scientifically rigorous proof.[3]

Meitner now continued to work on her own and in 1918 separated a long-lived residue from pitchblende. With a half-life of 32,000 years this isotope—protactinium 231—is the most stable form of the element, and credit for its discovery goes to the Berlin team. Hahn's war work allowed him only occasional visits to Berlin, so Meitner carried out the experiments largely alone. The programme had always been a collaboration with a mutually agreed goal, but its culmination was very much thanks to Meitner. Their work reporting their discovery of "a new long-lived radioactive element [that is the] mother substance of actinium" was published in June 1918 under Meitner's name, "after experiments carried out jointly with Dr Hahn".[4]

Meitner increasingly focused on beta decay. As Bohr's theory had anticipated, the production of protactinium is indeed the result of a beta decay from uranium. Today we know that this is the result of a positive beta particle being emitted. This positive analogue of the electron is known as a *positron*—an example of antimatter—but the concept of positrons and antimatter was yet for the future. In the early 1920s this process was interpreted as a kind of beta decay in reverse, where the negative beta is captured by the nucleus thereby lowering uranium's total charge by one.[5]

When the war ended, Hahn returned to Berlin where he and Meitner resumed their research collaboration. Their complementary expertise in radioactivity and chemistry made them a formidable pair. They were also technical innovators and built the first cloud chamber in Berlin. A cloud chamber is a device where electrically charged particles pass through supersaturated mist and ionise atoms in the vapour. Drops of liquid condensing on the ions create trails like those that reveal the passage of high-flying aircraft. Little wisps and threads of cloud form instantly around the ions produced by the radiation, and when illuminated the tracks stand out like motes in a sunbeam. The result is like having a telescope that can reveal what goes on in the subatomic world.

Hahn and Meitner used their cloud chamber to reveal the trails of alpha particles being slowed as they passed through the air. Their measurements produced a detailed catalogue of the alphas' ranges from many radioactive sources. These data enabled them to back-determine the sources of alpha particles being produced by mere traces of radioactive elements in complex samples. Thanks to this knowledge, they were confident that the alpha particle they had detected from pitchblende in 1917 had to be from protactinium and not from some other element.

Meitner continued her work on beta decay and measured the beta spectra of nearly all radioelements. She was increasingly being recognised as a leading scientist in her own right. She had maintained contacts with the Institute for Radium Research in Vienna, from which she got a supply of actinium in 1923. This enabled her to continue experiments into beta and gamma radiation produced in the cascade of radioactive elements from actinium to bismuth and lead. From 1922 to 1924 her studies on the "constitution of the inside of the nucleus of radioactive substances" were cited by the Austrian Academy of Sciences who awarded her the Lieben Prize.[6] In 1922 she was made a Privatdozent—essentially allowed to teach at the university—the first woman in Prussia to be so appointed.

Only in 1922 did Meitner start to publish regularly in her name alone, her work aimed at clarifying the patchwork of beta and gamma rays coming from uranium and other radioactive elements. By then it was clear that some elements emit beta electrons from the nucleus, some emit alpha particles, and some emit both. In 1924 Meitner and Hahn were nominated for the Nobel Prize in Chemistry, but the Committee chose not to award any prize in the subject that year.[7] Nevertheless, by the latter half of the 1920s, nearing fifty years of age and at the height of their powers, they were recognised by the international community of atomic scientists as leaders in the field.

Thanks to the work of Rutherford, Soddy, and, not least, Hahn and Meitner, in about fifteen war-interrupted years a phenomenological understanding of the energy streaming from uranium and a whole family of its radioactive descendants was in place. The role of alpha decay as the means to progress from uranium to lead or via actinium to bismuth was also established. The structure of the atomic nucleus and the reasons why some nuclei exhibit radioactivity were yet mysteries.

In France, where the radioactive journey had begun a quarter of a century before, attention at the Curie Institute had concentrated on building intense radioactive sources, not on the causes of the phenomenon. In Germany, Hahn and Meitner had led the way by identifying the pattern of radioactive transitions, but even now the Curie Institute had not focused attention on the big question: how is the nucleus built? This would soon change.

THE SUN KING—AND QUEEN

Some years ago, I was examining the notebooks Rutherford kept at the time when he discovered the atomic nucleus. Trapped between the pages was a slip of paper, folded in two, which looked as if it had lain there unopened for decades. The experience was like seeing a fossil emerge from a rock. A message written in the maestro's bold hand read: "Mme Curie thinks 6 grammes will be enough". There was no explanation of when this was written or what the 6 grammes was. As for Mme Curie, she needed no introduction.

After Pierre Curie's death in 1906, when he slipped and fell under a heavy horse-drawn cart, Marie was left to care for their intense daughter, Irène. In 1910 Marie began an affair with a married physicist Paul Langevin. The resulting scandal gave the misogynists of the French Academy of Sciences an excuse to refuse to allow her membership—not until 1979 would it admit a woman.

In 1914 Marie Curie inaugurated the Radium Institute in the Latin quarter of Paris. Four days after this, the First World War began. To protect the nation's supply of radium from German invasion, it was moved to Bordeaux. During the war, Marie and Irène put their scientific expertise to use with twenty ambulances—known as Les Petites

Curies—containing radiology equipment that X-rayed injured troops at Ypres and at the locations of several other battles. It's been estimated that a million men were treated, doubtless saving many lives. Meanwhile radium was being used by the military to provide soldiers with "see-in-the-dark" timepieces where the dials and hands were painted with radium that glowed. These watches became all the rage postwar.

So fascinated were the Curies by radium's mystery, they and many others failed to recognise its dangers. For example, at watch factories in the United States, teams of women applied radium to the dials by using brushes made from camel hair. To maintain the shape of the fine bristles the workers would mould them into points by licking the hairs. By 1925 several women died from "radium jaw"—a painful swelling of the upper and lower mandibles. By now Marie Curie too was increasingly laid low by bouts of illness—anaemia, overall fatigue, and general weakness—symptomatic of the cancer that would eventually kill her.

When the war ended, the Radium Institute was an empty shell. The Curies had never patented the production of radium, and funds for the project were hard to come by. However, when Marie Curie visited the United States in 1921, the resulting publicity and financial support reinvigorated the Institute, and she now prepared to hand over the running to Irène. Which is where Marie's former lover and now close friend, Paul Langevin, played a key role for both the Institute and the future of science.

Langevin was a scientific advisor to the French military and director of science at a Parisian technical school where he chanced upon Frédéric Joliot, an exceptionally skilled student. As a child Joliot had conducted scientific experiments in his parents' kitchen where he had taped to the wall photos of his French scientific heroes: Pierre and Marie Curie. Pierre Curie had also taught at Joliot's school.

At Langevin's urging, Joliot chose physics as his major subject at Paris's ESPCI (École Supérieure de Physique et de Chimie Industrielles). After a period in the military, and armed with a letter of recommendation from Langevin, Joliot marched confidently into Marie Curie's office at the Radium Institute on 16 December 1924, age twenty-four, and asked to become her assistant. The French scientific community was a clique, both professionally and socially, and hard to break into. The letter from Langevin must have helped as she agreed to take Joliot on.

Marie Curie's deputy, André-Louis Debierne, warned the young enthusiast that as the world already knew so much about radioactivity, there was little more to be done in the way of new research. But then chance intervened. The first week of 1925, Marie put Frédéric Joliot in the charge of Irène.

Three years older than Frédéric, Irène was one of few scientists worldwide who had trained in radiochemistry. The two were very different personalities. Irène—cold and intense—was the Crown Princess of France's scientific royalty; Frédéric, a charming extrovert, talented but with no obvious scientific pedigree, was the frog who had won her attention. What they had in common was a total dedication to science, a persistent determination not to be deterred by failure, and the drive to uncover nature's deepest truths. They fell in love and married in 1926.

Elitist French intellectuals viewed Frédéric Joliot as an opportunist, treating his decision to adopt the surname Joliot-Curie as evidence.[8] Nonetheless he and Irène were first-rate scientists. Initially they worked independently, Irène on alpha rays emitted by polonium and Frédéric on the electrochemistry of radioactive elements. In 1927 they started to work collaboratively and between 1927 and 1929 developed new techniques for separating polonium.

One outcome of Marie Curie's high-profile visit to the United States was that the Radium Institute received spent radon capsules that had been used in cancer treatment. Once the radon decayed it was no longer of use to the doctors, but its decay products included polonium 210, a more powerful source of alpha particles than radium. Having no further need for the radon capsules, medics in America, and increasingly from around the world, sent them to Paris, both as a means of disposal and as a tribute to Marie Curie's discovery of radium. As a result, the Radium Institute accumulated the world's largest source of polonium.

By 1931 the Joliot-Curies had purified a sample of polonium ten times more intense than any in existence. They now had access to the most powerful source of alpha particles in the world and the means to probe atomic nuclei with unrivalled clarity. Where previously Marie and Pierre Curie's vision had been focused more on the medical applications of radioactivity than its origins, now, with Irène and Frédéric joining forces, radioactivity would become a research tool. The second generation of France's royal family of science now prepared to take centre stage in the atomic drama. The stimulus would be a strange discovery made in Berlin.

HOW TO MISS A NOBEL PRIZE

In 1929 Berlin, physics professor Walther Bothe and his student Herbert Becker began a series of experiments probing light elements with alpha particles. When electric charges in atoms are disturbed, they emit electromagnetic radiation, the resulting spectrum giving information about the atomic structure. They reasoned that a similar phenomenon should happen with nuclei whose protons carry electric charge. Disturb the electrically charged nucleus with the impact of alpha particles, and gamma rays—very high frequency electromagnetic radiation—will be emitted.

They detected the photons with a Geiger counter. This is an excellent way to do it because when photons pass through the ionised gas inside the counter, the more intense the gamma rays are, the more pronounced the characteristic click of the counter becomes. The problem is to be sure that the photons came as a result of the alpha particles having hit the target and not directly from the radioactive source of the alphas itself. Meitner's experiments on the gamma radioactivity of various elements had found that radium is an intense emitter of gamma rays. Radium was also a convenient source of alpha particles, which had been ideal in experiments hitherto where the photons were irrelevant, but now could threaten the whole purpose. By contrast, polonium 210 emits alphas powerfully with very few photons. So, any photons detected by the Geiger counter with the polonium source would likely have arrived from the nuclear target, or in any event be much larger in intensity than the noise of any unwanted photons coming directly from the polonium source.

The problem was that polonium at that time was hard to come by, being available only as a product of radium decay, and radium itself was still in short supply. How Bothe and Becker got hold of their polonium isn't known, though with Meitner being present in the same city, and with her contacts with the Vienna Radium Institute having given her access to material for her own experiments, that might have been the way. In any event, they used polonium, which gave them a nice source of alpha particles with minimal gamma ray background.

Bothe and Becker were aware that the alpha particles would not have enough energy to penetrate the powerful electrical barrier surrounding the nuclei of heavy elements. Consequently, they focused on the lighter members of the periodic table, from the silvery metal lithium at number 3 to oxygen at 8 as well as magnesium, aluminium, and silver. They found gamma excitation with several elements, which was more or less what they had expected, except that in the case of beryllium, a metal which is the fourth lightest atomic element in the periodic table, they found two surprises. First, the intensity of what they measured as gamma rays was more than ten times larger than that of any other element. Second, the energy of the radiation was apparently larger than the laws of physics would allow, unless the beryllium nuclei were disintegrating. Yet the beryllium nucleus appeared to be remarkably solid as no protons were ejected. Fixed as they were on gamma rays, they reported the results of their successful enterprise. They commented on the anomaly for beryllium in December 1930 but gave it no special attention.

In France, however, the Joliot-Curies took note. By 1931, Irène and Frédéric's polonium store gave them a source of alpha particles ten times more intense than any other existing in the world. Armed with this product of years of purifying polonium, they put it to use to confront the beryllium conundrum.

And when the Joliot-Curies' beam of alphas blasted the target of beryllium, they too saw the mystery radiation. Their better experiment than the Germans' created further surprises. On 28 December 1931, Irène told the French Academy of Sciences that the radiation was even more intense than the Germans had found. Furthermore, she estimated its energy to be about three times that of the incident alpha particle. As to what it consisted of, they could not yet say, but, as it carried no electric charge, they, like the Germans, assumed that it consisted of high energy gamma rays.

They continued experimenting over New Year's Eve and into the start of January. They confirmed the radiation's presence and its power by placing paraffin wax, a soft solid made of hydrocarbons, downstream. Hydrocarbons contain atoms of hydrogen, and, sure enough, when the neutral radiation hit the wax, protons were ejected. They announced this discovery to the French Academy of Sciences on 18 January 1932.[9]

In retrospect it is astonishing that the French pair did not immediately realise that the radiation was not gamma rays but massive neutral particles—neutrons. Like a cue ball at snooker hitting the pack and ejecting one or more similar balls, the neutron had hit a target nucleus and ejected a proton. With this oversight they also missed the Nobel Prize that would have followed.

One reason for their failure was that they were unaware of Rutherford's 1920 Bakerian Lecture in which he had postulated the idea of a massive neutral partner to the proton. The French consequently knew only of gamma rays, like the Germans before them. Desperate to make a great scientific discovery of their own, to establish themselves and escape from the shadow of Irène's famous mother, the Joliot-Curies misinterpreted what they were seeing. Later, Rutherford met Frédéric Joliot and asked him, "Did you not realise that you had found those neutrons that I discussed in my Bakerian Lecture?" Joliot haughtily replied, "I never read your lecture. I thought it would be the usual display of oratory not of new ideas."

7
Il Papa

THE "MOST WORTHY" ENRICO FERMI

Among the scientists bunkered 10 miles from the atomic bomb test in New Mexico on 16 July 1945 was Italian Nobel laureate Enrico Fermi. Fermi's ability to estimate the magnitude of any physical phenomenon, making good approximate calculations with little or no actual data, is famous in the scientific community. What he did when the bomb exploded is typical and has become legendary.

A few seconds after the blast, Fermi stood up, tore a sheet of paper into small pieces, held them aloft and then threw them in the air. Seconds later, when the shock wave hit, the confetti was blown a few yards away. Fermi paced the distance across to where they had landed and then consulted some notes that he had made earlier. From this he estimated the force of the blast to have been equivalent to about 10,000 tonnes of TNT. A week later, detailed analysis of the Trinity test concluded the magnitude to have been about 18,000 tonnes, within a factor of two of what Fermi had estimated in less than a minute of the explosion. One of Fermi's biographers would later remark: "None of the scientists were surprised."[1]

Born in Rome at the turn of the twentieth century, Enrico Fermi was a child prodigy. By the time he started university in 1918 he had mastered all classical physics. What would become the two pillars of twentieth-century physics—relativity and quantum theory—were both very new and no one in Italy was yet teaching them. So, the teenaged Fermi taught himself.

Fermi, who would become one of the twentieth century's greatest physicists, was physically one of the smallest, at about 5 feet 3 inches, or 160 centimetres. His precocious talent came to the notice of Orso Corbino, a Sicilian and nationally rated physicist with an administrative and political genius. In 1922, shortly after Fermi graduated from the University of Pisa, this balding middle-aged operator with a bushy moustache became Fermi's mentor.

Corbino was born in 1876 near Catania and became a professor of physics in Rome. During the First World War he devoted his research to the war effort, which brought him to the attention of the military, industrial leaders, and politicians, who recognised his administrative skills and technical acumen. By 1920 he was a senator and in 1921 Italy's Minister of Public Education. In addition to these national responsibilities, he continued his work as a physics professor in Rome.

Corbino was distressed that no one in Italy was taking note of the extraordinary advances in relativity and quantum physics, and he was desperate to remedy this. Meanwhile Fermi, the student who had taught himself these subjects, probably understood them better than most of the physics faculty. In Italy, relativity theory was the province of mathematicians. By the time of his graduation in July 1922, Fermi was the only expert on relativity in the Italian physics community, so it was to his and Corbino's mutual fortune that at the very time when Corbino was fretting over how to remedy this national failing, Enrico Fermi arrived at his office on 28 October.

Corbino's dream was to restore the great Italian physics tradition symbolised by Galileo and Leonardo da Vinci. One of Corbino's great strengths was his judgement of talent, and he immediately saw in Fermi the means to achieving that renaissance.

In 1923, at much the same time that Fermi met Laura Capon, his future wife and the daughter of an admiral, Benito Mussolini made Corbino his Minister of Natural Economics. In a stroke, Fermi's innate genius was supplemented by a direct line to the heart of the Italian government. Corbino's first act was to arrange for Fermi to visit Germany where the likes of Werner Heisenberg and Wolfgang Pauli were developing the new theory of quantum mechanics. Fermi stayed for a year, learning how these ideas could revolutionise the understanding of atomic physics, and then returned to Italy well versed in the current state of the art. In 1925 Fermi made his first breakthrough by applying quantum theory to the statistical mechanics of thermodynamics.

Thermodynamics was a child of the First Industrial Revolution. Its laws linking mechanical work, energy, and heat were the foundation of nineteenth-century physics and chemistry. Thermodynamics deals with macroscopic quantities, describing the energy of a gas, for example, in terms of overall concepts such as temperature and pressure. The microscopic picture is one of the gas's individual molecules in motion, their kinetic energies collectively giving the phenomenon of heat such that the greater their motional agitation, the higher the temperature is. Conversely, were all molecules to have no kinetic energy and be at rest, the temperature would theoretically fall to some absolute minimum.

Not all molecules in the gas jiggle at the same rate, however. By chance some will be more agitated than the mean, others less so. The concept of temperature arises from a statistical distribution of the motions of individual molecules. A given temperature is linked to probability, the distribution of slow and fast being statistically spread. The microscopic theory underlying thermodynamics is known as statistical mechanics.

In statistical mechanics, the motions of the individual constituent molecules are controlled by the classical mechanics of Isaac Newton. This was the case until the cusp of the twentieth century, at which point the focus on atomic structure revealed that classical mechanics is not appropriate for the micro world. Instead, the new quantum theory was the rule, as applied so successfully by Niels Bohr in his description of the electron in the hydrogen atom. Fermi, who was fascinated by probability throughout his career, combined quantum theory and statistical mechanics to describe what would become known as the *Fermi gas*—the first of many concepts that would carry his name.

Quantum theory was born in 1900 when the German theorist Max Planck introduced the concept of a quantum "in an act of desperation" to reconcile the laws of thermodynamics and the experimental data on the electromagnetic radiation emerging from hot bodies.[2] Thermodynamics implied that the hotter the body, the greater the intensity of ultraviolet rays and X-rays would be. Instead, empirically the spectrum was found to peak at some frequency, dying away in the ultraviolet and higher frequencies. The frequency at which the intensity peaked indeed increased as the temperature grew, but there was always then a falling off in the ultraviolet region and beyond. Planck discovered that he could mathematically describe the data if he assumed that electromagnetic radiation is not continuous but is absorbed and emitted in discrete packets which he called *quanta*. The energy of a quantum is proportional to the frequency or colour of the radiation. Bohr took this concept over from electromagnetic radiation and applied it to the electron in his atomic model. By 1913, quantum theory looked complete.

However, Bohr's model—while successful for hydrogen with just one electron—ran into problems when extended to elements with several electrons in their atoms. The key insights to the solution came from the Austrian theorist Wolfgang Pauli. Pauli introduced the concept that an electron carrying electric charge also has magnetism with a duality where the north or the south pole can orient along or opposite to a magnetic field. This colloquially is known as an electron's spin, though this is just a picture as a point-like piece of charge literally spinning is unreal. More radical was Pauli's invention of the *exclusion principle* where no two electrons in an atom can occupy the same quantum state. This extended Bohr's picture beyond hydrogen, explained the periodic recurrence of chemical properties of the atomic elements, and would win Pauli the Nobel Prize in 1945.

Fermi saw in the exclusion principle the key ingredient for marrying quantum theory and the classical Newtonian theory of statistical mechanics. He extended Pauli's idea by applying it to electrons *outside* the atom, such as a gaseous plasma of electrons or when they are moving in a metal. All these electrons would have quantum energies and obey Pauli's principle.

Fermi constructed the relevant quantum equations of statistical mechanics and published them in 1926. Their implications for understanding previously baffling data in magnetism and electric currents in metals were immediately recognised. Today their applications extend to semiconductors, computers, MRIs, and much of modern technology. At the age of twenty-five he immediately became one of the world's leading theoretical physicists, and the sole Italian in that select group.

Corbino was now worried that Fermi would be offered a post at some prestigious European centre, so he quickly had a professorship created at Rome with benefits that Fermi would find irresistible. On 7 November 1926, Fermi was installed as Rome's professor of theoretical physics. Corbino's recommendation had included the prescient observation: "He is the best prepared and most worthy person to represent our country in the field of intense scientific activity that ranges the entire world".[3]

THE VIA PANISPERNA BOYS

Fermi's life was engrossed by physics, for which he had a childlike enthusiasm. His working day was fixed. He would be an early riser, work through the morning until lunch, and take a break for a couple of hours until the mid-afternoon. Then he would go back to calculating, which would continue through to the evening with a break for dinner. He would work until bedtime having little interest in socialising or cultural activities such as the theatre. His relaxations were outdoor activities such as hiking, climbing, or playing tennis, in all of which he was intensely competitive.

He was a natural leader, but initially in Rome he was on his own. Creating a school requires students, but the most prestigious physical science was engineering, with good career prospects; physics was not a popular choice. Corbino decided to encourage a few students to transfer from engineering to physics, by promoting the opportunity to learn from and work alongside the man he presented as the new da Vinci. While giving a lecture to second year engineering students, Corbino announced that he was recruiting a brilliant young scientist to the physics faculty and offered the students the chance to join in a field that was undergoing a revolution.

The plan worked. Two who made the move were Emilio Segrè, a future Nobel laureate, and Edoardo Amaldi, who postwar would become the founder and director general of CERN, the European Centre for Nuclear Research. The Rome physics centre was at 89A Via Panisperna, home of the Corbino family who lived at the top of a large house with three floors and a basement. The second floor contained a library and research laboratories, while classrooms were on the ground floor. The building was isolated from Rome's bustle by a high wall, with palm trees and a pond in an extensive garden giving it an air of calm.

Initially there were four in the group—Segrè, Amaldi, and Fermi himself, along with Franco Rasetti, a contemporary of Fermi who had known him as an undergraduate. Tightly knit, they learned from Fermi in what amounted to personal tutorials, and increasingly they all spent time together, hiking on weekends.

Segrè was quickly impressed and so certain he had made the right choice that he told another engineering student—a Sicilian named

Ettore Majorana—and encouraged him to put his talents to the test with Fermi. Majorana had a reputation in the engineering faculty as a prodigy, correcting lecturers' errors and, when challenged, spontaneously being able to complete proofs on the blackboard of half-given theorems, unrehearsed, from the power of his mind alone. A few months after Segrè transferred to the physics faculty, he encouraged Majorana to follow his example. Segrè explained that Fermi was a pioneer, that great things were happening, and that Ettore could be a part of it.

Fermi's breakthrough applying Pauli's exclusion principle to his formulation of statistical mechanics in 1926 had already marked him as a prodigy and one of the quantum pioneers. After that he applied the same approach to the cloud of electrons surrounding an atomic nucleus. When Majorana came to visit him in the autumn of 1927, Fermi was attempting to work out the consequences of this new theory for atomic electrons.[4] To do so he had been solving the equations day in and day out for about a week by manipulating an adding machine. After many thousands of operations, he had obtained a numerical table of results. This was what he was doing when Majorana arrived at his office in 89A.

Majorana asked Fermi to explain the problem and after their discussion he left without further comment. Within a few days, working at home alone, Majorana transformed the equations into a set that he could solve analytically. He then returned to the physics department, asked Fermi to show him his numerical results, and compared them with his own. Having verified they agreed, Majorana casually remarked to his fellow students: "Surprisingly Fermi has made no errors".[5]

This was not a comment on any perceived limitation of Fermi's powers, far from it. What Majorana was acknowledging was that in the labour-intensive numerical operations, Fermi had maintained perfect accuracy. Majorana must have been impressed as he decided Fermi was worthy of his attention and transferred from engineering to theoretical physics, making him the fifth of the Via Panisperna boys, as they became known. Meanwhile, Majorana's performance had impressed the existing boys immensely, as until then they had thought no one could rival Fermi.

With youthful zest they gave one another nicknames. Perhaps appropriate to Rome, ecclesiastical monikers were favoured. Fermi's infallibility made him Il Papa, the Pope; Franco Rasetti, Fermi's right-hand man, was the Cardinal Vicar, while Corbino who held the purse strings was Padreterno—The Heavenly Father, God Almighty.

The "boys" were very convivial, talking freely about politics, girls, and life. Fermi, a man with no airs and simple in manner, was always easily accessible but, outside of physics, reserved and not given to private confidences. When Majorana joined the group, he was soon named The Grand Inquisitor, in recognition of his profound scepticism, both of others and his own work. Like Fermi, Majorana was reserved and held back from the exuberance of the others.

The background to all their physics in the late 1920s was fascism under the dictator Benito Mussolini. For political convenience Fermi joined the fascist party claiming that politics meant little to him but explaining he would do what was necessary to be able to pursue physics without interference. The group meanwhile was becoming internationally recognised, the Via Panisperna laboratories in Rome now an institution of choice for international scholars. One of these, a young theorist named Gian-Carlo Wick, joined the boys. Wick was very anti-fascist and was prepared to say so. Fermi said he personally had no political views but preferred that there be no public expression of anti-fascism.

For its first two years, Fermi's group had focused on electrons and the outer reaches of the atom, but in 1929 Fermi told Corbino that he judged study of the atomic nucleus to be "the true field for the physics of tomorrow".[6] Corbino was so impressed that he promised the Italian Society for the Progression of Science in September 1929, "Italy will regain with honour its lost eminence." He repeated Fermi's assessment: "The only possibility of great discoveries [is] that we might be able to identify the internal [structure of the] nucleus of the atom".[7] By 1929 the state of knowledge about the nucleus was still limited: the nucleus contains the bulk of an atom's mass, and it contains protons, but otherwise its constituents and structure were a mystery.

Mussolini's government did not come through with money to enable Rome to develop "big" science, in the form of expensive machinery to blast the atomic nucleus. In the United States at the University of California, Berkeley, however, Ernest Lawrence was building the first cyclotron—a device to speed charged particles like protons around circular paths before whipping them off the track to smash into atoms, the goal being to explore the nucleus at high energies. Fermi decided to send his team to learn from those on the spot at major foreign labs. This included Cambridge and Berlin, where Rasetti went to see Meitner and learn from her how to build and operate a cloud chamber to detect particles.

Fermi meanwhile concentrated on theoretical aspects of nuclear physics. He felt certain they would have to modify the laws valid for the atom before obtaining a satisfactory theory of nuclear phenomena. The most obvious puzzle was: what holds large nuclei together? Gravitational force is much too feeble, while electric force—the only other then known—seemingly dictated that nuclei could not exist. For example, the positively charged nuclei of heavy atoms such as uranium attract large numbers of negatively charged electrons, but this creates a paradox for the nucleus itself: like charges repel so there must be some strong attractive force at work to hold the nucleus together. Fermi calculated that "nuclear bonds are a few million times greater than most chemical bonds".[8] He was correct but at this stage had no insight as to how these bonds could be formed.

Fermi was also the first to recognise the potential of another insight from Pauli, this time concerning beta decay.

Lise Meitner's careful measurements of beta decay had found the energy of the beta particles varied from one experiment to the next. This seemed contrary to the conservation of energy. Einstein's formula relating mass and energy, $E = mc^2$, implied that a radioactive nucleus with atomic mass m_1 contains an amount of energy E_1, and after emitting a beta particle the resulting nucleus will contain a fixed amount of energy determined by its mass, E_2. By the conservation of energy, one of the sacred principles of physics hitherto, the energy of the beta particle should therefore be equal to this difference, namely some fixed amount given by the difference $E_1 - E_2$. Yet Meitner clearly found that from one measurement to the next this energy varied, not by much admittedly but it was definitely not a fixed number.

In December 1930 Pauli wrote to her with the idea that beta decay involved the emission of not one but *two* particles, the electron and an electrically neutral particle that escaped detection but carried off some of the energy. She was about to attend a convention on radioactivity, in Tübingen. Pauli was unable to be there but asked Meitner to present his proposal along with his apologies in a letter addressed to "Dear Radioactive Ladies and Gentlemen".[9] He pointed out that the fixed energy therefore would be shared between the two, such that focusing on the one—the detectable beta particle—gave a spread of numbers. Fermi was one of very few at the convention who took the idea seriously. He dubbed this hypothetical particle the *neutrino* and locked the idea away in his mind, convinced by his natural intuition that this seemed right.

Fermi would later bring it to fruit, but not yet. His exploitation of Pauli's neutrino hypothesis would only be possible when the immediate question of how beta radioactivity occurs was answered, and this would first require understanding how the nucleus is constructed. The answer to that key question of nuclear structure had already been revealed, but no one yet realised it. The first to do so seems to have been Ettore Majorana.

ETTORE MAJORANA AND THE NEUTRON

Majorana would disappear in 1938 without a trace at just thirty-one years old, but for which he would undoubtedly have become a household name. This is how Fermi summarised Majorana's gifts and why he regarded him as the greatest theoretical physicist of the twentieth century: "If a problem has been proposed, no one in the world can resolve it better than Majorana".[10] During his brief life, Majorana had several groundbreaking insights, but limited his observations to notebooks, declining to publish even seminal breakthroughs that he regarded as trivial. In the opinion of Emilio Segrè, a future Nobel laureate, Majorana was mathematically "superior to all of us, and in some respects even to Fermi", which while impressive seems tame given Segrè's further comparison: "He was a prodigy... one couldn't know if Majorana would become a second Newton".[11] He published a mere nine articles in his career, and then only after insistent pressure from Fermi.

Early in 1932, when Majorana read Irène Joliot-Curie's account of the mystery radiation and her belief that it was gamma rays, he realised her mistake immediately. To Majorana, the idea that massless gamma rays could knock massive protons out of nuclei and eject them at tens of thousands of miles an hour was absurd. Even though he was unaware of Rutherford's speculations about a massive neutral "atom" in his 1920 Bakerian Lecture, Majorana had immediately understood what the data meant. The French "haven't understood anything", he told colleagues: "What fools! They have discovered the neutral proton and they do not know it."[12]

As Majorana explained, the phenomenon, as described by the Joliot-Curies, ran counter to sacred principles of the conservation of energy and momentum. The problem is this. Gamma rays are well known to be able to kick electrons out of atoms, but here they would be ejecting protons, nearly two thousand times heavier than electrons. To blast a proton out of its resting place in line with what the French duo saw would require each gamma ray photon to have about eight times the energy of the alpha particle that had initiated the sequence: alpha particle hits beryllium, liberating a "photon" which then impacts the target of paraffin wax, ejecting the protons. The idea that photons could emerge with eight times the energy supplied made no sense. However, if the culprit was a neutron, whose mass is like that of a proton, all could then be understood.

Fermi urged Majorana to publish this stunning observation, but Majorana regarded his remark as so obvious and trivial as to be of no consequence. Fermi was planning to attend a conference in Paris, at which the Joliot-Curies would be present, so he asked Majorana if he could at least report on his work there. Majorana reacted furiously: "I forbid you to mention these things that are so stupid. I don't want you going around discrediting me."[13] Fermi respected Majorana's request. So, not only had the Joliot-Curies missed the neutron, but Majorana failed to take the credit for identifying it.

8
In Bed for a Fortnight

Each day James Chadwick would walk or cycle from the Gothic quadrangles of Gonville and Caius College in Cambridge, where he was a Fellow, about a mile along King's Parade past the splendours of King's College chapel until he cut away into Free School Lane. Less than 3 metres wide, the lane is a canyon bordered by stone. On one side is the blank grey wall of Corpus Christi College, and on the other is a red brick building whose windows could have doubled for offices in a Northern mill town. Here was the main entrance to the Cavendish Laboratory, where during the 1920s Rutherford had attracted a team of talented young physicists with Chadwick as his deputy.

Although Chadwick had quit on his attempts to produce a neutron in 1928, the idea of this neutral constituent of the nucleus was never far from his mind, so he was well prepared when fortune struck on the morning of Friday 22 January 1932. He entered the Cavendish as usual, collected his mail, and took it to his office on the first floor. Opening his post, he discovered freshly arrived from Paris a copy of the French physics journal, *Comptes Rendus*, containing a paper by Irène Curie and Frédéric Joliot.[1]

As Chadwick absorbed their arguments, he had the same insight as Majorana: it was impossible for *massless* photons to have enough punch to kick protons out of paraffin wax, as the French thought. Instead, something with mass was required, having enough heft to do the job. Chadwick could hardly believe his luck as it dawned on him that Curie and Joliot had blundered in their interpretation of the experiment's results: they had proof of the neutron, the prize he'd been chasing for ten years, and had completely failed to realise it. Yet unaware of Majorana, Chadwick rightly suspected it would soon be obvious to many, not least the Joliot-Curies. To have any chance of succeeding in his historic quest, Chadwick had to act fast.

By the time he had read the paper for a second time, he had already decided how best he could repeat and improve their experiment to prove the neutron's existence beyond any doubt. This would require various pieces of scientific equipment, which he would have to get from the Cavendish store, and as a first step he would have to convince the director, Ernest Rutherford, that he was right.

JOSTLING THE CROWD

In 1932, the forty-year-old James Chadwick was at the height of his powers as a research physicist. His conclusion that the Joliot-Curies had made a fatal error was not based purely on intuition. Bothe and Becker's discovery in 1929 that beryllium emits intense radiation when hit by a beam of alpha particles had led Chadwick to assign the task of investigating this phenomenon to a student, Hugh Webster, a twenty-six-year-old physics graduate from Tasmania, one of a cohort of young Commonwealth scientists attracted to Cambridge by

Rutherford's presence. In the summer of 1931, Webster discovered that the intensity of the radiation was greater when it was emitted in the same direction as the incident jet of alpha particles than when emitted back upstream, in the opposite direction. For Chadwick this result provided a compelling hint that the radiation was not gamma rays, which like the illumination from a lightbulb should have beamed in all directions uniformly. Instead, the radiation's prominence when collimated with the alpha particle beam suggested, in Chadwick's judgment, that it consisted of massive particles, which had been ejected forwards from the beryllium by the impact of incoming alpha particles. That the Joliot-Curies' paper made no mention of this told Chadwick that they were unaware of the phenomenon.

The year before, Chadwick had correctly deduced that, in reality, these "gamma rays" were massive neutrons, but having made the right deduction, he was still blinded by the belief that his quarry was a composite of a proton and an electron. Based on that mistaken hypothesis, Chadwick had decided that Webster should use a cloud chamber. Chadwick's intuition was that the neutrons would fragment into protons and electrons and the passage of these electrically charged particles would be revealed as they traversed the chamber. Webster saw nothing, however. Chadwick was again frustrated in his quest—until he read the Joliot-Curies' paper in the third week of January 1932.

Chadwick had a regular 11:00 a.m. meeting with Rutherford where they would discuss work in progress in the laboratory along with any scientific news. That morning, 22 January, he told Rutherford about Curie and Joliot's paper.

Rutherford, who was as astonished as Chadwick, saw almost immediately that the protons could not have been ejected by neutral massless photons but instead by an electrically neutral particle with a mass like that of a proton. Chadwick later recalled Rutherford's "growing amazement" until he burst out "I don't believe it".[2] Rutherford, like Chadwick, believed their experimental observations but not their explanation. The two men were agreed: Curie and Joliot had misinterpreted their results.

It was possible that the Joliot-Curies had made an error in their measurements, but Chadwick and Rutherford judged that to be unlikely as the French duo were such excellent experimentalists. Convinced that the neutron was there to be found, Chadwick needed to prove it and—driven by the competitive urge that motivates many successful scientists—to do so before the French realised their mistake.

A problem was that unlike electrically charged particles, which are relatively easy to detect, a massive neutral particle would be all but invisible. Chadwick's idea was akin to H. G. Wells's story of *The Invisible Man*. Although the tale's hero could not be seen directly, he gave his presence away when he jostled the crowd in London's Oxford Street. For Chadwick, the crowd would be the nuclear protons whose electric charge makes them possible to detect. A proton will be ejected with almost the same speed as the incoming neutron, like one snooker ball hitting another. If an entire nucleus recoils, its speed will depend on its mass: the heavier it is, the slower it will recoil. From these measurements, Chadwick hoped to determine the mass of the neutral invader.

Chadwick reasoned that the radioactive blast of alpha particles upstream must have ejected these massive neutral particles from within the beryllium nuclei, which had then cannoned into the target atoms and ejected high-speed protons. Chadwick now explained to Rutherford how he could prove the fact by experiment and stressed the urgency if he was to do so before anyone else. A definitive experiment would require him to use as a target not just paraffin wax but samples of several elements to prove the phenomenon was a general truth. Graphite, which is a form of carbon, would provide him atoms of that element; nitrogen and oxygen gases extracted from the air could be pumped into a chamber to give him two other elements.[3]

"NEUTRON?"

The first piece of fortune for Chadwick was the timing because in January he had at last accumulated a good amount of polonium as an intense source of alpha particles. In Paris, as we have seen, the Curie Centre had amassed polonium from medical radon capsules, which were gifted from the United States; Chadwick was now the beneficiary of similar largesse. A former student of his, Norman Feather, had spent a year working on radioactivity in Baltimore and upon his return to Cambridge brought back three hundred used capsules of radon from a local hospital. During 1931 Chadwick extracted polonium-210 from these and by the start of 1932 had produced an ideal source of alpha particles. Even so, this amounted to just a thin film of the stuff which he spread over the surface of a silver disc about the size of a thumbnail. This he placed immediately behind a slightly larger disc of pure beryllium. Alpha particles from the polonium showered the beryllium, causing it to emit the neutral radiation.

To detect the outcome when the radiation hit a target, he used an ionisation chamber, a small device filled with gas that detects the passage of electrically charged particles, such as protons or atomic ions. On the side of the chamber facing the source was an opening, the size of a penny, through which the particles could pass. The amount of ionisation was recorded by electrical pulses displayed on an oscillograph. The size of the signal gave a measure of the energy of whatever had entered the chamber.

Nothing showed up until Chadwick repeated what the Joliot-Curies had done, putting a 2-millimetre sheet of paraffin wax in front of the window. Immediately the oscillograph burst into life showing that the neutral radiation from the beryllium was ejecting charged particles from the paraffin into the chamber. He then interposed sheets of aluminium foil between the wax and the chamber until there was no more signal. The way the signal died away confirmed that the charged particles were indeed protons, and from the thickness of the aluminium when the signal died completely, he was able to determine the protons' energy.

Thus far he had repeated the work of the Joliot-Curies. If protons are like snooker balls, Chadwick had confirmed that something was hitting them and ejecting them from the paraffin wax. Now he had to determine if that "something" was indeed a photon as the Joliot-Curies thought, or a heavy particle akin to a cue ball.

So began the novel phase of Chadwick's investigation. He removed the paraffin and inserted other elements in the way of the "cue ball". He used solids—lithium, beryllium, boron, carbon—and in every case the counter responded. Moreover, not just protons but even the whole atomic nucleus could be knocked into the chamber by the impact. Chadwick filled the chamber itself with gases—hydrogen, helium, nitrogen, oxygen, and argon. Each time the mystery rays gave the atoms of these elements a kick.

All this took an immense amount of effort and time. Each separate element had to be prepared and then a long series of readings taken. Anomalous results required him to check and repeat the measurements. Chadwick worked all day and then took a break for dinner in college. After dinner he would retrace his steps back to the Cavendish and continue taking measurements late into the night until exhausted. Cavendish folklore recalls him making this trek every day, while looking increasingly haggard. "Tired, Chadwick?" asked a colleague. "Not too tired to work" came the reply.[4]

Chadwick laboured day and night for about three weeks. To his delight he found that protons or atoms of the target elements were indeed ejected in every case, and always in the direction away from the beryllium source. Most important of all was the discovery that the energies of the recoiling protons and atoms were much too large if the mystery radiation consisted of gamma rays—massless photons.

He was left with two alternatives to explain what he was finding. First, that he gave up the fundamental principle of the conservation of energy and momentum; this would have been so radical as to have overthrown more than a century of science in a stroke. His second option was to maintain this sacred principle and to suppose that the mystery radiation consisted of massive particles. By careful measurements of the recoil from different atoms, he was able to show the mass of an incoming neutral particle to be very similar to that of the proton itself.

Everything about this was consistent with the neutron that Rutherford had discussed in his Bakerian Lecture in 1920. One further property that Rutherford had predicted was that it would be extremely difficult for neutrons to interact with matter. This also was consistent with what Chadwick was observing because when he placed centimetres of lead in the path of the radiation, it had little measurable effect on the intensity, showing that lead was effectively transparent to these particles. Protons, for example, cannot penetrate even a millimetre of lead as they are absorbed by the electrical forces within the solid; neutrons, by contrast, were predicted by Rutherford to have no such impediment, and the ability of the neutral radiation to pass through lead clinched his prediction.

At the end of that intense pursuit, Chadwick had established the neutron as the third basic constituent of matter, jointly with the proton and the electron. Less than a month after first reading the paper by the couple he always called the "Curie-Joliots", Chadwick sent a brief report of his completed work to the editor of *Nature* on 17 February. In the interim he had averaged perhaps three hours of sleep daily, including the weekends. In the cautious nature of science at that time, he called his report "Possible existence of a neutron", but as he said later, there was no doubt whatever in his mind as to what he had discovered, or he would never have made the claim.[5]

Ten years earlier, a young Russian physicist at the Cavendish, Peter Kapitza, had founded a club open to all members of the laboratory who were prepared to discuss their work openly and informally. Members met in college rooms after dinner on Tuesdays, and rumours began to spread around the laboratory that Chadwick would announce a big discovery at its 302nd meeting of 23 February 1932.

The venue was a room in Nevile's Court at Trinity College. The seventeenth-century quadrangle is bordered by the grandeur of Wren

Library to its west, and the college hall to the east. Its north and south sides contain two floors of rooms sited above covered colonnades. At over 60 metres long, it takes about one third of a second for sound to travel one full length and echo back from the stone wall at the far end. Three centuries earlier, Trinity College's most singular alumnus, Isaac Newton, made his first estimate of the speed of sound there, by clapping his hands and timing the echoes with a small pendulum—about 3.5 centimetres long—whose swing corresponded exactly to the time between the clap and the echo.[6] Now in 1932, in a room above those colonnades, members of the Kapitza Club gathered to hear the most momentous news about atomic structure since Rutherford's discovery of the nucleus itself. Beyond the rumours they had yet no details other than that some new understanding of the mysterious nucleus was to be announced.

It was a cold February evening; the wind having blown across the flatlands of the Fens all the way from the deep freeze of central Europe. A coal fire warmed the room, its curtains pulled across leaded windows. Armchairs and a sofa hosted about twenty attendees, and a small table at the front awaited Kapitza and the speaker.

In the college hall, Kapitza had wined and dined Chadwick into a "very mellow mood" and then accompanied the exhausted scientist the short distance to the meeting room. One of those present was a research student and later acclaimed author, C. P. Snow, who recalled it to have been "one of the shortest accounts ever made of a scientific discovery". Chadwick summarised: "We've found it. It's an uncharged particle which is part of the mass of the nucleus. It can move freely through matter as Ernest Rutherford said it would. I propose it be known as the neutron." Brief and to the point, he ended with the request "Now I want to be chloroformed and put to bed for a fortnight".[7] But first he completed the formality of signing the club records against the title "Neutron?", the query hinting that full confirmation was yet to come (Figure 5). He continued to make measurements, assisted by Feather, and by June was able to announce in his full-length report to the Royal Society, "The existence of a neutron", without any "possible" or "?".[8]

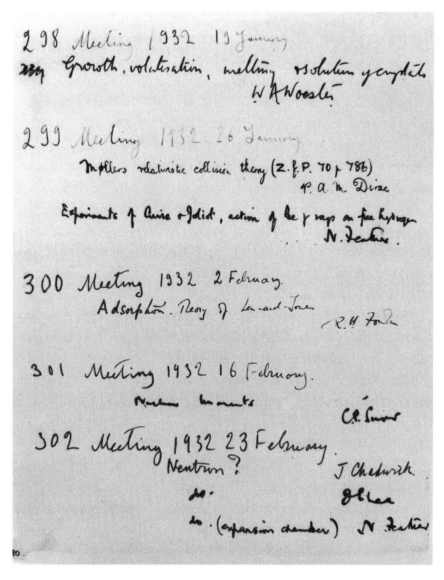

Figure 5. Neutron discovery: list of Kapitza Club talks, including those by N. Feather on 26 January and by J. Chadwick on 23 February 1932. (Credit A. Brown and Churchill College Cambridge.)

The caution advertised at the time of the Kapitza Club and in the title of Chadwick's first paper hid his undoubted confidence in his results. Club members had surely been primed as they had heard Feather mention the "Curie-Joliot" paper in a brief commentary after the main talk three weeks earlier.[9] Furthermore, Chadwick had invited to the meeting an old friend from his Manchester days, J. G. Crowther, the distinguished science correspondent of the *Manchester Guardian*. Crowther described the news as "one of the most fundamental discoveries in physics, [which] ranks with the discovery of the electron, the proton and the X-rays. It is hailed as the most important achievement in experimental physics since Lord Rutherford demonstrated the nuclei [sic] structure of the atom in 1911 [sic]". Crowther judged that Chadwick now joined "Bohr and Moseley as the most brilliant of Lord Rutherford's Manchester pupils".[10]

Apart from his paean to Chadwick, Crowther's report gave a profound insight into Chadwick's perception of what he had discovered: this was no fundamental particle but an amalgam of proton and electron, as Rutherford had long believed. For Chadwick, in Crowther's words, the neutron represented "the first step in the evolution of matter, the first step in the building up of the common materials of everyday life out of the primeval electrons and protons".[11]

Chadwick's discovery of the neutron was so important that it is regarded as the moment when the science of nuclear physics began. At the time, however, few outside the rarefied world of physics gave it much significance. It was hard enough to visualise an atom, let alone its nucleus. The neutron was both far beyond vision and of no apparent use, although the *Manchester Guardian* asserted practical applications would "doubtless be discovered before long". *The Times*, however, assured readers that discovery of the neutron and nuclear research in general "would make no difference".[12] No one yet realised that the neutron would light the nuclear fire. Far from Chadwick's vision of it as the "first step in the evolution of matter", the neutron would become key to unleashing forces able to destroy life on Earth.

9
Moonshine

RUTHERFORD'S AMBITION

The Cavendish scientists' attempts to reach the nuclei of heavy elements were being hampered by the rule of nature that like charges repel: alpha particles produced by natural radioactivity were too easily flung aside by the electric forces surrounding the atomic nucleus. Rutherford was all too aware of this limitation, which had troubled him for some time. In 1927 he had addressed it in his Presidential Address to the Royal Society. He said: "It has long been my ambition to have available for study a copious supply of atoms and electrons which have an individual energy far transcending that of the alpha and beta-particles from radioactive bodies".[1] By 1932, even Chadwick's highly intense polonium source couldn't break through this electrical shield. A way to create an atomic cannon of high energy particles had to be found.

The simplest ionised atom is hydrogen, which when stripped of its single electron leaves a proton. The idea of ionising hydrogen and then subjecting the protons to electric forces was talked about by members of Rutherford's team as a possible way to accelerate them. The proton is of course positively charged and so as a nuclear probe would be subject to the same repulsive forces as are alpha particles. However, the proton only carries half as much charge as an alpha particle, so the repulsion is less. The greatest benefit is that protons can be made available in large quantities—by ionising hydrogen atoms—and then given kinetic energy by electric forces. Just 1 milliamp of electric current in the form of high-speed protons would produce more particles than 100 kilogrammes of radium or polonium could. By focusing on protons and accelerating them, Rutherford's "ambition" might be satisfied.

The faster protons are moving when they encounter an atom, the greater is the chance of some reaching their goal: the nucleus itself. In the Cavendish Laboratory, heavy equipment to that end was built in collaboration with the industrial giant Metropolitan-Vickers. This included high-voltage machinery making potential differences exceeding half a million volts, typical of a lightning flash.

The principle of energy conservation implied that the electrical energy provided by high-voltage machinery can be converted into the corresponding amount of kinetic energy of the protons. The standard theory implied that many millions of volts would be needed for those protons to have enough power to penetrate the electrical shield and reach the nuclei of heavy elements. Banks of capacitors were mounted atop one another all the way from the floor of a laboratory to its ceiling in the hope of them cumulatively building up that vast number of volts. Hopes were dashed as sparks flew, the charge neutralising as electric currents suddenly streamed through the air. The nuclei of the heavier atomic elements seemed destined to be forever out of reach.

The breakthrough initially came from George Gamow, a Russian émigré and theorist then working at the University of Göttingen. Quantum *theory*, the natural philosophy of the atomic scale, had been accepted for two decades, but its laws of motion—quantum *mechanics*—were only formulated in 1925. These extensions of Newton's classical mechanics into the atomic world immediately proved remarkably successful when applied to the behaviour of electrons. In 1928, Gamow applied the new theory for the first time to nuclear physics. He showed that quantum mechanics explains how heavy elements such as uranium can eject positively charged alpha particles when Newton's classical mechanics would have implied this to have violated the principle of energy conservation and been impossible. Published in a German journal, Gamow's paper was initially read only by a small number of theorists specialising in the new quantum mechanics.[2]

In 1928, one of Rutherford's experimental assistants, John Cockcroft, learned of Gamow's result and had a flash of inspiration. Cockcroft had a degree in electrical engineering from Manchester and one in mathematics from Cambridge. He had been working under Rutherford at the Cavendish since 1924 and been stimulated by Rutherford's dream of energetic particles. Cockcroft imagined the process of alpha radioactivity in reverse, where it would be akin to a positively charged particle—the alpha particle—invading and successfully penetrating the electric shield of a heavy nucleus. What is good for alpha particles in theory should be good for protons too. Gamow's use of the newly developed quantum mechanics showed theoretically that penetrating the nuclear electric shield would be easier than originally thought. Instead of having to energise a beam of protons with many millions of volts to blast into a heavy nucleus, as classical mechanics would have predicted, a few hundred thousand volts would be sufficient to get them deep into the atomic realm where quantum mechanics takes over. Thanks to quantum mechanics, exploration of heavy nuclei appeared to Cockcroft to be a practical proposition.

When Cockcroft did the calculation, however, he found there was some fine print in the quantum contract: quantum mechanics makes predictions of probabilities, not certainties. While its prediction that a few hundred thousand volts could make a proton breach the barrier was good news, the downside was that there was less than one in a million chance that a specific proton would win the lottery. To have a real chance of success, many billions of protons would be needed.

Cockcroft and another of Rutherford's assistants, an unassuming young Irishman named Ernest Walton, spent the next three years designing, building, and testing their *proton accelerator*. The whole project was a tour de force, not least of the electromagnetic force. First, to make the protons for the beam they had to strip the electrons from atoms of hydrogen gas; this required 40,000 volts to pull the negatively charged electrons and positive protons in opposite directions. The protons would then be fed into a pipe where they would be accelerated by a potential of over 500,000 volts. While this was straightforward on paper, there were immediate problems such as how to have a 40,000-volt transformer making the proton beam very near to a 500,000-volt one that is needed to accelerate them. This involved immense amounts of insulation with the ever-present threat of receiving electrical shocks or even electrocution.

The protons would be accelerated in a 1-metre-long vertical tube, emptied of air. The 500,000-volt potential difference from top to bottom would hurl the protons down the tube until they emerged at the bottom with their maximum speed. At this they would fly into a chamber where a target would be placed.

Collisions between protons and target atoms would give signals to be recorded by scintillations on a zinc sulphide screen. The problem of human contact now became very relevant. Cockcroft or Walton would have to be sitting near to the target while looking at the screen through a microscope, which meant that they would be in danger of electrocution by 500,000 volts as well as threatened by possibly intense radiation from the collisions. So, for protection they built a small hut, lined with lead, in whose cramped confines they would control the experiments and make their observations. The whole device was about 4 metres tall, too high for a normal office, so it was installed in a disused lecture hall adjacent to the main Cavendish Laboratory.

The story of the atomic nucleus had until now involved groups of one or two individuals working in small rooms, their apparatus cobbled together from bits and pieces found in laboratory drawers supplemented by glass tubes prepared by an assistant. Cockcroft and Walton's machine signalled a new era in science. A transition from small to big experimental apparatus was beginning.

It took eighteen months for them to complete this, which brought them to May 1930. Only now were they ready to test the machine, first at 50,000 volts, then 100,000 volts, stepping up gradually, they hoped, to 300,000 volts and beyond. There were many occasions where short circuits caused sparks, vacuum tubes sprung leaks, along with other disasters, but fortunately nothing broke that couldn't be fixed. By 1931 the main challenges were to be able to reach higher voltages and to make more intense beams. The goal of irradiating atomic nuclei and seeing what happens was still an indetermined way off.

LAWRENCE OF AMERICA

Rutherford's plea in 1927 for a source of high energy particles inspired physicists and engineers both in the United Kingdom and in America, notably the twenty-seven-year-old Ernest Lawrence, who arrived in Berkeley, California, as an associate professor of physics. Lawrence read an article by a Norwegian engineer, Rolf Widerøe, which showed how to accelerate a particle to moderate energies by a series of small pushes from a series of relatively low accelerating voltages. Lawrence had found his scientific calling: developing particle accelerators.

Widerøe's device, which he built in 1928 at the Technical University of Aachen, consisted of a sequence of separate metal cylinders in an evacuated tube. Within each cylinder there were no electric fields, and the particles would coast along. Across the gaps between cylinders, however, Widerøe set up electric fields with alternating voltages, switching between positive and negative values. He cleverly matched the oscillation frequency so that when a particle emerged from a cylinder it would receive a kick not a brake. This accelerated the particle every time it crossed a gap between one cylinder and the next, the effect compounding until the particle had reached the desired energy.

Reaching the high energies needed to attack the nucleus, however, would have required a vast number of repetitions and been utterly impractical. Were it not for this feature, Widerøe's device would have been a brilliant fulfilment of Rutherford's ambition, but in its original form it had little to offer for nuclear physics. Enter Ernest Lawrence. Lawrence's brilliant insight was to use a magnetic field to steer the particles into a circular orbit. This would in effect enable each particle to cross the same acceleration gap many times rather than traversing many gaps once only, as in Widerøe's creation.

Lawrence's design was to have a hollow circular metal cylinder, about the diameter of two hands, which he cut into two identical halves known as dees, after their shape. He then separated the two dees with their straight sides a couple of centimetres apart. The 2-centimetre separation would form the gap across which the particles would be accelerated. He injected particles from the radioactive source into one of the dees through an opening at the side; magnets above and below then steered each particle around a circular orbit in the plane of the container.

By charging one of the dees positive and the other one negative, Lawrence produced an electric field across the gap between them. If the direction of the force was right, the particle would be accelerated across the gap, after which the magnetic field would swing the particle through a semicircle, bringing it to the gap at the other end of the dee's straight edge. On this, the second half of the circuit, the particle would cross the gap but in the opposite direction. To be accelerated both times, the direction of the electric field must also be swapped, which Lawrence achieved by switching the positive and negative voltages to each of the dees so that their polarities were reversed.

The way he achieved this was by exploiting a beautiful implication of Isaac Newton's equations of motion. When a particle gains energy and goes faster, as here, centrifugal force makes its arc larger, which means it has further to travel before reaching the next gap. What Newton's equations imply is that the length of the path increases in exact proportion to the speed, which has a magical consequence: the ratio of length to speed—in other words, the time—stays the same. So, the time taken to complete half a circuit remains fixed whatever the particle's energy.[3] All Lawrence needed to do was to match the frequency at which the electric field switched with that of a particle completing half a circuit. Particles from a radioactive source were then injected through an opening and spiralled outwards to the edge, eventually emerging with greatly increased energy.

Lawrence had the idea of combining magnetic swings and electric pushes, along with the key property of "equal time" per circuit in 1929, but he seems to have done no more with it for a year. It was only when he took on a research student named M. Stanley Livingston, in 1930, that he decided to work this up into a realistic project. They did so together, and colleagues jokingly referred to the device as a cyclotron—a name that has stuck. They built a device about 4.5 inches, some 12 centimetres, in diameter, which successfully accelerated particles to a kinetic energy of about 80,000 electron volts. By the summer of 1931 they had built an 11-inch (28-centimetres) cyclotron which accelerated the particles to an energy of 1,000,000 electron volts, or 1 MeV.

In parallel to Cockcroft and Walton across the Atlantic, Lawrence was spawning what would become known as "big science", for although his cyclotron itself was relatively small, the associated apparatus needed to inject particles, the magnets to steer them, targets for them to probe, and the detectors to record the results were large. By 1939 his enterprise had grown to such an extent that a cyclotron of diameter 1.5 metres with its associated paraphernalia occupied its own dedicated building on the University of California campus, known as the Radiation Laboratory. The "Rad Lab" became a Mecca for the research and development of particle acceleration, and that year Lawrence won the Nobel Prize for his invention of the cyclotron. However, back in 1931, he had been so focused on the technology that the team at Berkeley neglected to apply it. The baton to be the first to blast atomic nuclei with man-made equipment was by 1932 unwittingly passed back to Cambridge, where Cockcroft and Walton were still having trouble refining their construction into an efficient working machine.

THE ATOM SMASHER

By late 1931, Cockcroft and Walton had been working for nearly three years on their machine, refining it, improving it, with the goal of reaching 800,000 volts. One of the problems was that they needed high voltages to have any chance of the protons penetrating the nucleus, but as atoms are mostly empty space, most protons would miss the nucleus anyway. To optimise their chances required the large numbers of protons to be tightly focused into an intense beam. What Cockcroft and Walton discovered, unfortunately, was that as the voltage went up, the beam got fuzzier. How to resolve these two conflicting features, namely the need for high energy and the need for an intense beam, took up more time. Rutherford began to get impatient.

Chadwick's discovery of the neutron in February 1932 had created huge excitement in the Cavendish. His breakthrough had climaxed in just twenty days and been achieved in the old-school manner of table-top apparatus and human ingenuity, whereas Cockcroft and Walton had been toiling for years without any positive results. On Wednesday 13 April Rutherford came into the lab to see what progress they were making. A series of events put him in a bad mood. First, he hung his wet coat on a live terminal and got an electric shock. Then he lit his pipe; it was full of dry tobacco which went off like a volcano. He summoned Cockcroft and Walton and grumbled about them wasting time. He told them to "stop messing about" and to "do something with their protons".[4]

So, they installed a zinc sulphide screen to detect—they hoped—scintillation flashes from charged particles even though their machine was not yet up to their planned peak power. Their first choice of a target was lithium, a silvery white metal, which they placed directly in the path of the beam. By the morning of Thursday 14 April, all was ready.

That morning Cockcroft had business in the university and so Walton began the experiment himself. He sat on a small stool in the cramped hut adjacent to the target and the detector, with his eye to the microscope pointed at the zinc sulphide screen. He turned on the beam and was astonished to see the screen immediately light up. There were not just one or two flashes per second but hundreds, far too many to count. He turned the beam off and the screen went blank. He was young and inexperienced and thought that this was an indication of alpha particles, but he was not sure. An internal phone system had recently been installed in the Cavendish and so he phoned Cockcroft.

The more experienced scientist came over to see the flashes for himself. He thought that they were indeed alpha particles but to confirm the fact said they should call Rutherford.

Now Rutherford arrived. A big man, middle aged and no longer supple, he eventually managed to squeeze into the small hut and sit on the stool. Cockcroft and Walton were a short distance away, by the electric control switch, and responded to Rutherford's sonorous commands: "Switch off the protons! Switch them back on! Increase the voltage!" and so it went on for several minutes.

After a few repetitions—beam on, scintillations visible; beam off, scintillations cease—Rutherford emerged, convinced the signals were genuine proof of alpha particles. "I should know. I was there at the birth of the alpha particle and have been observing them ever since" he assured them. Rutherford called up Chadwick who came in to view the sensation. Chadwick agreed: "They're alphas".[5]

Rutherford described the sight as "the most beautiful in the world". A simple piece of arithmetic explains what had happened and shows the reason for his excitement. A nucleus of a lithium atom consists of three protons and four neutrons. Add one proton to that—provided by Cockcroft and Walton's machine—and you have four of each or, which is what Rutherford immediately saw, two of each, twice: in other words, two alpha particles, the nuclei of helium atoms. The protons had not just hit the lithium nuclei; they had split them clean in two. Lithium is one of three primordial elements, along with hydrogen and helium, made in the Big Bang. After 13.7 billion years, it had become the first to be cleft by human ingenuity.

Less than 300,000 volts were required to achieve this, which also endorsed Gamow's application of quantum mechanics to nuclear physics. His work had been a theory of how, thanks to quantum mechanics, alpha particles can tunnel out of a nucleus. Cockcroft had applied the logic in reverse to estimate how much energy would be needed for protons to get *into* the nucleus. The resulting splitting of lithium showed that this had worked as anticipated. So, thanks to Gamow, their experiment implicitly also explained how alpha radioactivity occurs.

This was the first time that apparatus built by humans had reached into the atomic nucleus and, moreover, split it. The term *atom smasher* was about to enter the lexicon of science. But not yet, as Rutherford quickly realised the threat and opportunity.

Cockcroft and Walton immediately drafted a letter to the editor of *Nature*. Once published, the world would know that splitting atoms needs much less energy in practice than previously thought. There were nine days before the next issue of *Nature* would appear, and Rutherford ordered that the laboratory stay open at night so that they could work around the clock bombarding as many elements as possible. Cockcroft, Walton, Chadwick, and Rutherford were the only ones who knew and, until the letter appeared in *Nature*, that is how it had to remain.[6]

There were scientific riches to be mined; Rutherford wanted his team to capitalise, fast. One colleague said Rutherford "treated the atom like a coconut shy at the village fair: he would throw things at it to see which bits dropped out".[7] Chadwick's discovery of the neutron heralded a nuclear wonderland of clusters of neutrons and protons, capable of being rearranged, alpha particles chipped off, and nuclear energy released. Lithium had been split in two; what unexpected delights might be revealed if protons bombarded other elements? Rutherford ordered total secrecy. None of the four were to breathe a word about what had happened, not even to colleagues at the Cavendish. This focus on secrecy was unique in the whole of Rutherford's career.

They continued the experiments by reducing the voltage and as expected the rate of the flashes slowed. Even when the voltage was reduced to 125,000 volts, two or three flashes a minute occurred. This showed that the proton acceleration technique was remarkably effective, the reason being they were producing about one hundred million protons a second, which meant that some got through and hit the target nucleus even when only moderately accelerated.

Next, they measured the energies of the alpha particles and found that these were about 8,000,000 electron volts. This is astonishing as their production had come from the impact of a proton which was only receiving 125,000 volts of applied electrical energy. That a relatively feeble proton could liberate such vast energy from the nucleus is a manifestation of Einstein's equivalence of energy and mass. The interpretation is that energy latent in the mass of the lithium nucleus, which was being used to hold it together, is released when it splits into lighter components.

The atomic masses of lithium and alpha particles relative to the proton were each known, and from that the amount of the energy, their mc^2, could be computed. When these were each checked against the measurements that Cockcroft and Walton were finding for the splitting of lithium into two alpha particles, the numbers balanced. This was the first direct proof of Einstein's formula in a laboratory experiment.

Up to this point Cockcroft and Walton had been detecting the alphas by their scintillations on the zinc sulphide screen; now they replaced the screen with a cloud chamber, which made trails of these charged particles visible. By the evening of Friday 15 April, they had not only succeeded but had taken photographs of tracks clearly caused by alpha particles as they traversed the cloud chamber. The range of those tracks were telltales for their energies; the results agreed precisely with what they had calculated from $E = mc^2$. Two separate methods of detection had now produced the same conclusions: absolute proof that the lithium nucleus was being split into two alphas had been achieved.

The letter that had been sent to *Nature* arrived after that week's deadline, so was not to be published for another week. This gave them more time to carry on experiments before the world heard about their discovery. They replaced the lithium target with one of boron. Boron consists of five protons and six neutrons and so when hit by a proton gives six of each. This equals three alpha particles, and, indeed, their cloud chamber showed three alpha particle tracks emerging after the proton hit the target. Experiments where a proton hit fluorine (nine protons and ten neutrons) produced a single alpha particle (two and two) while converting the fluorine into oxygen (eight and eight). So, within two months of Chadwick having discovered the neutron, Cockcroft and Wilson's experiments with protons were confirming that nuclear clusters of these two near-twins could be rearranged by man-made intervention, converting one element into another.

On 28 April Rutherford took Cockcroft and Walton along to the Royal Society to be present at a meeting celebrating Chadwick's discovery. There was a triumphant speech about that seminal breakthrough, followed by questions from the audience. When this was over, instead of formally closing the meeting, Rutherford stayed at the podium. There was a dramatic pause. Then, after a few seconds he announced "a new and wide field of research" had opened up and he wished to introduce the two young men who had in recent weeks disintegrated lithium, boron, fluorine, and aluminium.[8] Cockcroft and Walton were brought to the front of the hall, and the audience gave them a standing ovation.

One of the most remarkable features of their achievement was not just that disintegration of the atomic nucleus released some of the energy that had been latent within the original, but that it liberated much more energy than it had taken to speed the protons. At first sight this seemed to be the holy grail, a means to extract useful energy from atoms. Unfortunately, however, the process was so inefficient—for example, fewer than one in a million protons was likely to hit an atomic nucleus—that a year later, in an address to the British Association for the Advancement of Science, Rutherford would famously remark: "Anyone who looks for a source of power in the transformation of atoms is talking moonshine".[9]

10
The Magicians

FERMI'S LIST

Fermi ranked physicists in divisions. Those in the second or third division "do their best but do not go very far", he declared, after placing some of the twentieth century's Nobel laureates in these lower rankings. His first division, which included himself along with the likes of Einstein, Rutherford, and the remaining laureates, consisted of "those of high standing who come to discoveries of great importance, fundamental for the development of science". Above even this premier division, Fermi had a tiny group that he dubbed "the magicians". This paramount category contained a mere handful of "geniuses like Galileo and Newton". Although Fermi's professional colleagues called him the

Pope, he had one team member whom he thought even cleverer than him, and into this godlike magic circle Fermi now placed Ettore Majorana, the reclusive young Sicilian who had joined him at the Via Panisperna laboratory in 1927.[1]

Chadwick's discovery that neutrons can be kicked out of a nucleus suggested that they existed inside them—atomic nuclei are tight clumps of protons and neutrons. Following Chadwick's success, "before Easter" 1932 and before Cockcroft and Walton's experiments had confirmed beyond doubt the content of the atomic nucleus, Ettore Majorana had developed a theory of lightweight nuclei made of protons and neutrons (he called them "neutral protons").[2] He knew that according to quantum mechanics, swapping the positions of two particles like a do-si-do in dancing can cause a mutual attraction, entangling them in a form of quantum knot. When the electrons in each of two hydrogen atoms perform this routine, a diatomic molecule—in chemical notation H_2—results. Majorana now applied this property to build a theory of atomic nuclei. Instead of electrons, the performers now are the constituent neutrons or protons. As the switching of electrons binds hydrogen atoms together, so will the exchange of nucleons (the collective name for neutron or proton) bind a nucleus.

This binding comes at a price: the atomic mass of an isotope, measured for example by deflection of a beam of isotopic ions in a magnetic field, is empirically less than the sum of the masses of its individual nucleons. The difference between the total mass of the neutrons and protons when free and the mass of the nucleus when they're all bound within is because some energy has been used to bind the nucleons together. The amount of this binding energy is determined, as always, by the formula $E = mc^2$. It is like an energy tax imposed on the nucleons for their accommodation in the nuclear prison.

The larger the magnitude of the binding energy, the smaller the mass of the total cluster will be. Smaller mass means smaller energy overall and greater stability. To liberate one or more nucleons, the relevant fraction of binding energy would have to be supplied or some configuration of larger binding energy—greater stability—be accessible. The actual stability of a nucleus is often determined by the size of its binding energy relative to those of other isotopes of the element, or of near neighbours in the periodic table of elements, which can be reached by alpha or beta decay. Alternatively, a nucleus may split into two or more fragments when energy is supplied by an impact, as in Cockcroft and Walton's experiments.

Majorana now worked out for the lightest elements of the periodic table which combinations of neutrons and protons had the largest binding energies and hence greatest stability. Key to his analysis would be that these nucleons followed the rules of quantum mechanics analogous to the way that electrons had done so in Bohr's model of the electronic atom.

Bohr had theorised that those atomic electrons occupy discrete energy states. Majorana now intuited that similar rules would apply for neutrons and protons within an atomic nucleus. Where we used the analogy of rungs on a ladder of energy when describing Bohr's electronic atom, so we can do for the neutrons and protons here. The only difference is that there are now two ladders: one for the protons and one for the neutrons.

Like electrons, protons and neutrons are magnetic and can orient their magnetic moments—north and south poles in effect—in either of two orientations. Quantum mechanics dictates that if the first proton (or neutron) on a rung has its north pole in some direction, a second proton (or neutron) must orient its south pole that way. By contrast, an individual proton and a neutron, being on separate quantum ladders, have no such restriction. This is Pauli's exclusion principle at work: the north-pole-up quantum state already being occupied would force the second identical member to adopt a north-pole-down orientation. The north and south poles being oriented this way causes their mutual magnetism to cancel, minimising their total energy.

Two protons in magnetic counterpoint and two neutrons likewise constitute four items in all on the lowest energy rung with their magnetic energies also being minimised. This lowest rung has in the jargon become saturated. The resulting state of low energy is very stable. It is the cluster known as the alpha particle—the nucleus of a helium atom.

Having successfully described the structure and features of helium and explained how pairing two protons and two neutrons gave a very stable combination, Majorana now considered the effects of other combinations of protons and neutrons in his equations. This led him to explain the pattern of isotopes for the lightest members in the periodic table of elements. Three, four, and five protons brought him respectively to lithium, beryllium, and boron, and when he included neutrons as well, a hierarchy of isotopes of varying relative stabilities emerged. Majorana's theory successfully identified which isotopes are stable and explained the propensity of others to stabilise through radioactive decay.[3]

When Majorana reached carbon and nitrogen, the sixth and seventh elements, the calculations became difficult. Convinced he was stupid, Majorana fell into depression, deciding his theory was worthless. Nonetheless, when he talked about this work in a seminar at Rome, Fermi and other first-rate physicists in the audience thought Majorana's ideas were groundbreaking and urged him to publish. Majorana would have none of it, as he regarded his work to be incomplete. Fermi had been invited to take part in the physics conference in July that year in Paris and had chosen as his subject the properties of the atomic nucleus. He asked Majorana for permission at least to mention his ideas on nuclear forces, but in a replay of his reticence with the neutron, Majorana again refused.

MAJORANA'S EXCHANGE WITH HEISENBERG

The young German theorist Werner Heisenberg was one of the founders of quantum mechanics—the extension of Newton's classical dynamics of the macro world consisting of planets, apples, and billiard balls, to the microscopic world of atoms, electrons, and atomic nuclei. In 1927 he proposed his eponymous uncertainty relation setting limits on how precisely the position and speed of a particle can be simultaneously determined. Heisenberg's uncertainty principle has for a century been recognised as one of the fundamental axioms of quantum mechanics.

At the Paris conference, Fermi learned that Heisenberg had built a theory of the nucleus similar to what Majorana had developed. Upon his return to Rome, Fermi immediately outlined Heisenberg's ideas and explained that they were less well-grounded than Majorana's. Once again, he urged Majorana to publish, and as before, Majorana refused: "Heisenberg has already said everything that could possibly be said", he responded, "and, furthermore, he has probably said too much".[4]

Like Majorana, Heisenberg had correctly identified the neutron as a key partner in building atomic nuclei, but like Rutherford he had wrongly perceived it to be a tightly bound coupling of electron and proton. According to Heisenberg, nuclear forces were the result of neutrons and protons exchanging these electrons. In his theory, a proton upon accepting an electron became a collective of a proton and an electron—the neutron as perceived by Heisenberg—while conversely this collective neutron could shed its electron and become a proton. Majorana, correctly and uniquely, regarded the neutron as a basic particle on equal footing to a proton. For him the exchange of electric charge had nothing to do with an electron but was the result of a fundamental neutron and proton swapping positions in their entirety. Hence Majorana's opinion that Heisenberg, by extending Rutherford's flawed hypothesis to build a theory of nuclear forces involving the exchange of electrons, "has said too much".

It was then that Fermi succeeded in persuading Majorana to visit Heisenberg at the University of Leipzig. Majorana arrived in Germany on 19 January 1933. Apart from a few weeks in March and April when the University was on vacation and he visited Niels Bohr in Copenhagen, Majorana was in Leipzig for six months, where in addition to the stimulation of discussing theoretical physics with the quantum pioneer, he had a political awakening as he experienced with horror the Nazis seizing power. Heisenberg, who made him welcome, quickly recognised that Majorana had identified shortcomings in his own theory and had independently built a consistent description of nuclear forces. Heisenberg succeeded where Fermi had not in convincing Majorana to write a paper about his theory of nuclear structure. Within a month, he had done so.

His paper, the German title of which translates as "On nuclear theory", was completed on 3 March 1933. Majorana began by reviewing Heisenberg's attempt, which "was guided by an analogy between the normal neutral hydrogen atom and the neutron, the latter being supposed—as commonly accepted—to be composed of a proton and an electron. The use of such an analogy is difficult to justify".[5] Majorana identified two problems with this.

First, in Heisenberg's model, protons and (composite) neutrons would result in nuclear goo, whereas Majorana's fundamental neutron generates structure. This is a consequence of Pauli's exclusion principle.

Protons are *fermions*, having one half of the basic unit of quantum angular momentum, or spin, and as such are subject to Pauli's edict: two such particles cannot occupy the same quantum state. Majorana's neutron also has spin one half and is subject to the same exclusions. If the lowest energy states are already occupied, further protons and neutrons must go into higher energy states as more complex nuclei are constructed; but for this exclusion they could all condense into the lowest energy state, leading to nuclear goo rather than structure.

Majorana's theory, which explained nuclear stability, critically relied upon Pauli's principle applying to both protons and to neutrons. In other words, the neutron like the proton must carry spin of one half and be no more or less fundamental than its electrically neutral twin. In Majorana's words "This comes directly from empirical properties of nuclei, and we cannot give them up".[6]

This is in stark contrast to Heisenberg's model. His "neutron" consisted of a proton and an electron, each of which have spin one half.

As two halves make a whole, his electron and proton composite would carry an integer amount of spin. If this were the nature of the neutron, the particle would evade Pauli's principle and lead to instability of the nucleus. Heisenberg, surprisingly, seems not to have realised this.

Majorana identified a second problem in Heisenberg's conception: by Heisenberg's own fundamental principle of quantum uncertainty, the energies of the nuclear constituents are utterly incompatible with those required to confine an electron within the minute size of an atomic nucleus. Even leaving aside that there is no mechanism known that could cause the conventional hydrogen atom to collapse such that its electron ends up tightly contained with the dimensions of the proton, these other problems empirically undermined Heisenberg's theory. By contrast, they were circumvented in Majorana's formulation, which corrected and empirically improved Heisenberg's theory. Heisenberg "much appreciated the work".[7]

Majorana's paper describes his theory and highlights the role of the Pauli exclusion principle. It does not give examples, whereas the memories of his Via Panisperna colleagues imply he did so in Rome when he originally outlined his ideas and inspired Fermi. The fact that the particles exchanged their spatial positions but not their magnetic orientations was a key feature of Majorana's theory and dictated how the protons and neutrons aggregated into specific energy levels, or *shells* in the jargon. Pauli's exclusion principle limited the number in each shell. The lowest energy state can accept at most two protons and two neutrons giving a maximum of four in the shell. Thus, hydrogen has isotopes with one neutron (deuteron, the nuclear seed of "heavy hydrogen" or deuterium) or two neutrons (tritium) but not three (quadium).[8] A third neutron, required to form quadium, would be forced to a higher energy level, which makes the resulting isotope highly unstable, effectively nonexistent.

Two protons in different quantum states could exist alone, in principle, but their mutual electrical repulsion negates this. Add one neutron, however, and the additional strong attraction stabilises the system to give the isotope of helium, helium-3. Add another neutron, giving a total of two protons and two neutrons, and the lowest energy shell is now saturated. The result is the highly stable alpha particle, the nucleus of helium-4. Add more neutrons and, again, these are forced to higher energy levels leading to unstable forms.

It was Majorana's focus on the positional exchange of fundamental protons and neutrons that empirically elevated his theory above Heisenberg's. It clarified why the alpha particle became so stable and does not decay to leave two deuterons; Heisenberg's theory had no explanation of this.

There is no doubt that these two remarkable young theorists respected one another highly. In a letter home, Majorana wrote "Heisenberg's company is unique."[9] As for Heisenberg, he understood Majorana both as a physicist and as a person: "He came as a young Italian physicist to Leipzig. He was a very brilliant man and at the same time a very nervous type of man. He did excellent work [but] was always extremely pessimistic about physics. I tried always to induce him to write papers and so he did finally write a very good paper."[10] That "very good paper" was Majorana's explanation of nuclear stability. There could be no better testimonial to Majorana's genius than the way that throughout his life Heisenberg—about to win the Nobel Prize that year for his work founding quantum mechanics—repeatedly stressed the importance of Majorana's nuclear theory, which replaced Heisenberg's misconstrued attempt.

HUNGARIAN RHAPSODY

It was in September 1933, a few months after Majorana's theory of nuclear forces appeared, that Rutherford made his much-quoted remark opining the idea of obtaining power from the atomic nucleus to be "talking moonshine". Nearly eighteen months had passed since the discovery of the neutron, and since Cockcroft and Walton split a lithium atom by bombarding it with beams of protons. The British

Association for the Advancement of Science in its 103rd year held its annual meeting in Leicester from 6 to 13 September 1933. On 11 September there was a "Discussion on atomic transmutation" led by "Rt Hon Lord Rutherford of Nelson O.M., F.R.S."

Rutherford's 10:00 a.m. presentation, lasting half an hour, was followed by Cockcroft and Walton speaking for a similar amount of time about their experiments using high-velocity protons. The next talk, by a young Australian at the Cavendish Laboratory, Mark Oliphant, commented that heavy elements—in particular lead and uranium—are not disintegrated appreciably when bombarded by protons, unlike their lighter counterparts. Finally, completing the session, another Cavendish scientist, Philip Dee, described how trails of the debris in cloud chambers had been photographed to confirm that when lithium is hit by a proton it splits into two alpha particles.

There is no mention of Rutherford's remark about moonshine.[11] Under Rutherford's name in the index, however, is reference to *The Times* of 12 September. Newspaper reporters had questioned Rutherford following his talk, and it was in response that he made his famous, and unscripted, remark. The precise words may have contained nuance which the media did not follow, for there are varying reports of what was actually said.

What the media reported has become folklore: "Energy produced by the breaking down of the atom is a very poor kind of thing. Anyone who expects a source of power from the transformation of these atoms is talking moonshine." I suspect that nearer the truth are claims that he hedged by adding, "with the means at present at our disposal and with our present knowledge".[12]

The Hungarian physicist Leo Szilard read the dramatic media version in his morning newspaper on 12 September. He was in London, staying at the Imperial Hotel in Bloomsbury. The oft repeated story, which Szilard was never slow to embellish, is that as he exited the front door of the Imperial Hotel and prepared to cross Southampton Row to reach Russell Square, he had his flash of inspiration.

Destroyer of Worlds

Leo Szilard was a thirty-five-year-old Jewish physicist from Hungary, who had been working in Berlin until he fled following the Nazis taking power. A squat man with thick curly hair, Szilard had a "powerful ego and invulnerable egocentricity".[13] Never understated, his self-confidence had led him as a PhD student in Berlin to ingratiate himself with Einstein.

At that stage Szilard's mainstream was research in thermodynamics, but his sideline was invention. During ten years in Berlin, he applied for twenty-nine patents, several with Einstein as partner, mainly applications of thermodynamics to refrigeration. At some point during the 1920s Szilard became interested in nuclear physics, however, and in 1929 independently of Lawrence came up with the basic idea of the cyclotron—and patented it. In 1932, following the discovery of the neutron, Szilard began to dream that nuclear energy might one day enable humans "not only to leave the earth but also the solar system".[14]

The Nazi takeover in Germany led Szilard to flee to Britain. His income from patents was substantial enough to pay his way, which is how by 12 September 1933 he was living at the Imperial Hotel in London when he read that Ernest Rutherford did not share his optimistic opinion on the opportunities for nuclear power.

The dramatic narrative, as Szilard told it, is that as he stood at the traffic light that morning, waiting for the stoplight to change so that he could cross the road, the sequence of red, to yellow, to green, somehow simulated the thought of a neutron impacting on an atomic nucleus, releasing both some energy and also another neutron, which could then continue the sequence in a chain reaction.

This oft-told tale has some flaws, however. If you are waiting as a pedestrian for the traffic lights to change to enable you to cross the road, the light does not sequence from red to yellow to green as this would enable the traffic to set off from a stationary situation when you could have crossed already. And as the first fully automatic pedestrian cross light in London was yet to be installed (at Piccadilly Circus in 1937), the story of Szilard and the crosswalk in 1933 may have been false memory enhanced by his subsequent experiences of American life where such signals were already common.[15] Even had the thought struck him after he had reached the far side, and watched the lights now changing to release the momentarily stalled energetic traffic, we are left with the question: why did this sequence of lights stimulate the idea not just of a chain reaction, which has some plausibility though perhaps coloured after the event, but with neutrons?

Szilard never explained this, but the fact that when he patented the idea he specifically associated this with releasing energy from a light element like lithium, suggests that the thought process covered not just Rutherford's remark but also what Cockcroft and Walton had already done: splitting lithium, albeit with protons. Recall what they had achieved.

Lithium consists of three protons and four neutrons. When struck by a proton, a lithium nucleus splits into two alpha particles, each of which consists of two neutrons and two protons (Figure 6a). Suppose we replace the incident proton by Chadwick's neutron. The lithium nucleus will split as before but this time into a single alpha particle and something else. The "something else" is key.

In Cockcroft and Walton's case this "something else" had itself been an alpha particle, consisting of two neutrons and two protons. But now we have replaced the incident proton by a neutron, so we are one proton deficient and have an extra neutron. There is no stable nucleus containing one proton and three neutrons to play the role of the second alpha particle in their experiment; instead, the result will be a nucleus of tritium—that is one proton and two neutrons—and an isolated neutron (Figure 6b).

So the result is that a single neutron has split lithium-7 to produce an alpha particle, tritium, and also a single neutron: we have split the nucleus by bringing in a neutron, and we have effectively recovered that neutron afterwards. The idea that this "secondary" neutron could now repeat the process, were it to hit another nucleus of lithium, might have been in Szilard's mind.

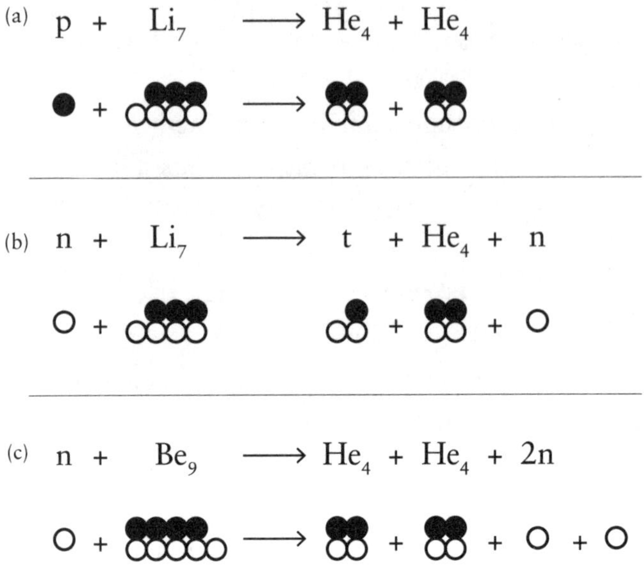

(a) $p + Li_7 \longrightarrow He_4 + He_4$

(b) $n + Li_7 \longrightarrow t + He_4 + n$

(c) $n + Be_9 \longrightarrow He_4 + He_4 + 2n$

Figure 6. The results of several atomic collisions: solid circles represent protons and open circles represent neutrons. (Formulas 6a and 6b from Close, *Trinity*, p. 427.)

Unfortunately, in the case of lithium this secondary neutron is too feeble to break up further nuclei; a chain reaction is in practice impossible. His patent focused on beryllium, the next element in the periodic table, consisting of four protons and five neutrons. It seems that he assumed the neutron would split beryllium into two alpha particles and shed two secondary neutrons (Figure 6c). If that were the case, one neutron incident would have produced two secondaries, and the process would multiply exponentially: two breeding four, which breed eight, which then breed sixteen, and so on. This seems too good to be true, and indeed so it is, as such a process takes up rather than releases energy.[16]

This lack of energy happens to be true for all light elements, which undermined Szilard's idea. Seemingly unaware of this, by March 1934 Szilard had developed his idea enough to file a patent application. With his usual chutzpah, he then sought an audience with Rutherford who agreed to meet him briefly on the first Monday in June. It was quickly apparent that the idea of a chain reaction was not new to Rutherford, but Szilard carried on regardless while making errors that Rutherford was not slow to point out.

When Rutherford heard that Szilard was patenting his thinking on the subject, he was horrified as this went so much against the convention that no basic science should be anyone's preserve. Szilard recalled, "I was thrown out of Rutherford's office." Szilard's patent was at that stage more imagined than real. He first offered the idea to the War Office who turned him down as they saw "no reason to keep the specification secret".[17] Finally, he signed the ideas over to the British Admiralty.

While the flaws in his proposal to release energy from light elements by a chain reaction were obvious to Rutherford and Chadwick, and possibly others, nobody seems to have applied Szilard's reasoning to the heaviest elements in the periodic table, such as uranium. His idea of a chain reaction induced by neutrons for the moment laid fallow.

INTERLUDE
The Birth of Nuclear Physics

1933 Solvay Conference

Back row: E. Henriot, F. Perrin, F. Joliot, W. Heisenberg, H. A. Kramers, E. Stahel, E. Fermi, E. T. S. Walton, P. A. M. Dirac, P. Debye, N. F. Mott, B. Cabrera, G. Gamow, M. S. Rosenblum, W. Bothe, P. Blackett, J. Errera, Ed. Bauer, W. Pauli, J. E. Verschaffelt, M. Cosyns, E. Herzen, J. D. Cockcroft, C. D. Ellis, R. Peierls, Aug. Piccard, E. O. Lawrence, L. Rosenfeld

Front row: E. Schrödinger, Mlle I. Joliot, N. Bohr, A. Joffé, Mme. Curie, P. Langevin, O. W. Richardson, Lord Rutherford, Th. De Donder, M. de Broglie, L. de Broglie, Mlle. L. Meitner, J. Chadwick

Absent: A. Einstein and Ch.-Eug. Guye (From US National Archives.)

The year 1933 was singular both in the history of nuclear physics and for the world.

In Germany the Nazis had come to power. An ironic consequence of the Nazis' racial laws persecuting nonethnic Germans was the migration of many leading Jewish scientists to the United Kingdom and North America, which enabled them to contribute to the development of nuclear science and the defeat of fascism. The most famous, Albert Einstein, left in 1933; many others followed his example throughout the 1930s. So began the sequence of events that would lead to the Second World War and its culmination with the explosive detonation of a nuclear weapon.

Lise Meitner was one scientist affected. In September 1933, she found herself forbidden to lecture at the University of Berlin because of her Jewish background, but as an Austrian citizen she was allowed to continue research at the Kaiser Wilhelm Institute in the same city. When her sister's son, Otto Frisch, was forced out of his position as a physicist at the University of Hamburg, he left Germany to live in the United Kingdom. There Frisch and Rudolf Peierls, another German Jewish émigré, would later make the breakthrough that would inspire development of an atomic bomb.

Along with Marie and Irène Curie, Meitner completes a female trinity of great women physicists of the early twentieth century. Following the discovery of the neutron, all three were invited in October 1933 to a special convocation in Brussels of the world's leading scientists.

Founded by Belgian industrialist Ernest Solvay in 1911, these sporadic invitation-only conferences brought together leaders in science to discuss the great issues of the day. Three of the early conferences have become famous in the history of atomic physics. The first, in 1911, focused on the nature and origin of radioactivity; the fifth in 1927 discussed the newly formulated quantum mechanics, which would eventually underpin much of nuclear science; the seventh conference from 22 to 29 October 1933 was dedicated to the structure and properties of the atomic nucleus. The resulting Solvay Conference of 1933 is today regarded as the pivotal moment when, after a long gestation, nuclear physics was born.

At the time of the conference, Irène Joliot-Curie was upset. She felt that she had some proprietary interest in the neutron and that Chadwick had invaded her territory. Irène and Frédéric, whose Curie heritage led jealous rivals to insinuate they were beneficiaries of nepotism, were desperate for a great discovery of their own.

Hardly had their failure to recognise the neutron sunk in when Carl Anderson in the United States reported his sighting of a positron—the antimatter form of an electron, carrying positive charge but with the same mass—passing through his cloud chamber which he had exposed to cosmic rays. That antimatter should exist in the form of a positively charged analogue of the electron was a prediction of the new quantum mechanics, so its discovery was another profound consequence of this radical description of the subatomic world. Like many other physicists, upon hearing the news the Joliot-Curies had immediately looked through their old cloud-chamber photographs from their experiments on radioactivity. To their frustration they discovered that their data had contained both negatively charged electrons and positively charged positrons. Having failed to recognise the neutron, they had also missed the positron.

Frédéric's announcement at the conference of their discovery that nuclear radioactivity emits positrons was interesting, nonetheless, and well received by the audience even if it felt to him and Irène at the time to be little more than a consolation prize. Irène now believed that in some features of their work they had made their longed-for breakthrough, and she proudly used the stage of the Solvay Conference to announce the fact. Unfortunately, the stress of having been pipped in the race for the neutron, and the positron too, had affected her judgment.

Beta decay in the form of negatively charged electrons emitted from atomic nuclei was by then well measured, not least by Lise Meitner, so following Anderson's discovery it was natural to suppose that positrons might also be produced in some radioactive decays. At Solvay, after Frédéric had confirmed the phenomenon of positive beta radioactivity, Irène went further. She told the distinguished gathering that when she irradiated aluminium with a beam of alpha particles, she detected radiation of positive particles, which she believed to be the first evidence of positrons produced in an induced nuclear reaction. She argued that some protons had converted into neutrons and positrons, and that the proton could be a composite of a neutron and a positron.

Even had her experimental claims been accepted, this suggestion was a step too far. Rutherford had believed a neutron was a composite of a proton and an electron; now Irène Joliot-Curie was suggesting that a proton could be composed of a neutron and the positron. The neutron and proton began to appear like matryoshka dolls. Where would the composition end?

Irène's moment of hoped-for redemption almost immediately disappeared when Lise Meitner weighed in against the French pair. Meitner, who was widely recognised as the world leader in the measurement of beta decays, announced that working in Berlin in 1932 she had conducted the same experiment that Irène Curie had just reported and that, although she had seen both positrons and neutrons, she insisted she had been unable to find a single example of a neutron accompanied by a positron "at the same time".[1]

Meitner was highly respected for her careful precision and the august body of scientists sided with her. Like many shy individuals, in company Meitner could be blunt and abrasive when discussing science. Her innocent remark, made in the calm logic of an analytic conference but with her forceful style, was psychologically devastating for Irène.

Immediately following Meitner's intervention, Francis Perrin, a theoretician from the Collège de France, suggested there might be two successive emissions, first of a neutron and then of "a positive electron with the intermediate formation of an unstable nucleus ([phosphorus-30] in the case of aluminium); this nucleus shows, in short, a radioactivity by positive electrons".[2] Perrin's vision began with the target nucleus of aluminium, which consists of thirteen protons and fourteen neutrons. The Joliot-Curies' addition of an alpha particle to the mix temporarily supplied two more protons and two more neutrons; however, the collision chipped off a single neutron from the nucleus, leaving a cluster of fifteen neutrons and fifteen protons. This group of thirty is a radioactive isotope of phosphorus, which with its thirty constituents would conventionally be called phosphorus-30. The phosphorus then decays by positron emission.

Perrin's remark was quickly overtaken by a debate between Bohr and Pauli on whether there was a strong analogy between the emission of positrons seen by the Joliot-Curies and the spontaneous emission of (negative) beta particles. This in turn led to a discussion of the nature of the neutron and the proton: is the proton composed of a neutron and a positron, as the Joliot-Curies believed, or are Rutherford and Chadwick correct that the neutron is made of a proton and an electron? The intervention by Perrin, at that stage still a relatively minor figure in theoretical physics compared to Bohr and Pauli, was forgotten.

That the neutron and proton are both equally fundamental constituents of atomic nuclei was the theme of a major theoretical talk at the conference by Heisenberg on nuclear forces. He repeatedly emphasised Majorana's theory—saying "according to Majorana", "as Majorana has stressed", "we will adopt, as did Majorana"—only occasionally mentioning his own contribution in an understated way.[3] Years later, Niels Bohr, the Danish Nobel laureate and father of atomic theory, wrote an account of the conference. In it, Bohr recalled this "most weighty report by Heisenberg" as the moment when science agreed that the neutron is a fundamental partner, a sibling to the proton, and not some composition of a proton and electron.[4] Heisenberg's paean to Majorana at the conference, and implicit denial of his own contribution, played a key role in establishing the neutron to be as fundamental a particle as the proton. For Irène Joliot-Curie this was a further body blow.

By the end of the 1933 Solvay Conference, a picture of the atomic nucleus had been agreed, and the basic theories of radioactivity outlined. Implicit to all this was confirmation of the neutron as a basic particle. As Heisenberg always generously admitted, the model of the nucleus built from protons and "neutral protons" (neutrons) was the progeny of Ettore Majorana.

Thirty-six years after the first glimpse of nuclear energy, twenty-one after identifying the atomic nucleus as a lump of electric charge, the origins of radioactivity and the structure of the atomic nucleus were understood. Liberating some of that nuclear energy by brute force had begun, but the challenge of releasing it fast enough to produce useful power remained. Much of this knowledge had come from individual brilliance, supplemented with communications by letter and sporadic one-on-one visits. The 1933 Solvay Conference was the first time most of the world's leading nuclear explorers had gathered for several days to share individual experiences, to compare, contrast, and debate what they understood. It marked the dawn of a new era.

The scientists returned to their institutions with a clear vision of how to proceed. Decades later, the distinguished nuclear theorist Hans Bethe would describe the annus mirabilis of 1932 and the ensuing Solvay Conference thus: everything that happened before this time was the pre-history of nuclear physics; everything that happened after is the history of nuclear physics.[5]

PART III

Releasing the Nuclear Genie
1933–1939

11
Fermi Explains Beta Radioactivity

The neutron was the legacy of alpha decay, and the Solvay Conference the legacy of the neutron. The legacy of beta decay and the Solvay Conference would be another electrically neutral particle: the neutrino.

In 1933, the neutrino was a hypothetical new particle, proposed by Austrian theorist Wolfgang Pauli to explain the energy emitted in beta decay. Today physicists posit new particles with abandon—quarks, the Higgs boson, supersymmetry particles, dark matter particles—but in 1933 this was not the case. The menu of matter's fundamental bricks back then appeared to contain just the electron, proton, and neutron. The emergence of the neutron had increased the previous basic pieces by 50 per cent, but it had been anticipated on theoretical grounds. And by 1933 there was also powerful empirical evidence: when alpha particles from natural radioactivity impact atoms of beryllium, neutrons are released as Chadwick had demonstrated. Thus was the neutron the legacy of alpha decay.

At the Solvay Conference the mysteries of beta decay were confronted. There was no theory of the phenomenon. This was in stark contrast to alpha decay, where George Gamow had shown that quantum mechanics enables tight bundles of pairs of protons and neutrons—alpha particles—to escape the grip of the strong nuclear force and then to be ejected rapidly by electrical repulsion between the now remote positively charged nucleus and the positively charged alpha. The alpha particles were clearly pre-existing constituents of a heavy nucleus. By contrast, there was no idea where or how beta particles originated.

That beta particles are electrons emerging in *nuclear* radioactivity, but otherwise identical to those that lurk in the outer regions of Bohr's atom, was by the time of the Solvay Conference agreed. The first theory of beta decay was inspired by what occurred at the Solvay Conference and came from Enrico Fermi. The story of how and why begins four years earlier when Fermi first decided that nuclear physics was the new frontier for his Rome group to attack.

NUCLEAR ELECTRONS

After Benito Mussolini and the fascists took control in Italy in the 1920s, the nation's society of leading scholars and artists—the influential Accademia dei Lincei—took stands that the dictator disliked. Il Duce decided to create an academy where cultured heroes would glorify Italy and fascism and, also, be superior to the Accademia dei Lincei. In March 1929 the first thirty members of Mussolini's invention—Reale Accademia d'Italia—included one physicist: Enrico Fermi.

Although party membership was not required of academicians, it was expected. Fermi seemed prepared to play Mussolini's game, letting his name and prestige be used, possibly as a price for ensured government funding of his subject. In any event, in joining the party he seems to have chosen the pragmatic course, which appeared to have been a guiding principle in both his politics and science.

Senator Orso Corbino, ever politically astute, had announced that nuclear physics would be the key to Italy's future and in 1931 encouraged Fermi to organise an international conference in Rome to be held at the Royal Academy. Corbino's hope was that by bringing frontier thinkers into contact with Fermi's group, the conference would kick-start the Rome nuclear physics programme. It would also show the political leaders how world scientific stars were flocking to Rome which they now saw as an important player in the field. Corbino was remarkably successful: Mussolini attended the inaugural session held on 11 October 1931.

Attendees at the conference included Marie Curie, Werner Heisenberg, Niels Bohr, Lise Meitner, and Wolfgang Pauli. These scientists formed a caucus in investigating the emerging mystery of beta radioactivity. Curie had helped start the saga long ago with the discovery of radioactive elements that had led to the appearance of alpha and beta rays. Once these were established as *nuclear* radioactivity, the first thought was that beta particles existed there already, like alpha particles. One person missing from Rome was Ernest Rutherford, whose decade-old idea of the yet undiscovered neutron as an electron-proton composite was at first glance an obvious way to explain not only nuclear masses and isotopes but also beta decay. However, there was a big theoretical problem: Heisenberg's uncertainty principle, a fundamental axiom of quantum mechanics, was a death knell for the idea that electrons can exist within a nucleus.

Relative to the size of an atomic nucleus, the atom is extensive. Heisenberg's principle showed that the kinetic energy of electrons within *atomic* dimensions is completely consistent with his uncertainty principle, but there was no way that electrons with the energies observed in beta decay could be contained within the narrow confines of the *nucleus*. This was no mere trifle either, more like a factor of over one hundred in error. Something was fundamentally wrong, and the conclusion was that beta electrons could not have pre-existed within the nucleus. Some other explanation of their appearance was called for.

Adding to the mystery was Meitner's observation that the energy carried by a beta particle seemed to disagree with the principle of energy conservation, which had inspired Pauli's proposal that in beta decay not only is an electron emitted but also a lightweight electrically neutral, ghostly particle—the *neutrino*. He mentioned his idea at the Rome conference, but to little enthusiasm. An invisible particle invented to solve one problem, for which there was no other need nor any indication of how the concept could ever be proved, seemed doomed by Occam's razor.

Niels Bohr, in desperation, suggested that energy conservation be abandoned at nuclear dimensions entirely. This was received with even less enthusiasm than Pauli's new particle. Like a politician of one party, who when asked an awkward question, replies, "What would the other party have done?", Bohr had no answer when asked what dynamics would replace the principle or how abandonment of energy conservation in the nucleus would connect with all the evidence that this sacred principle applies at atomic and larger distances.

That was the situation in 1931. Following discovery of the neutron and new forms of radioactivity, the Solvay Conference in 1933 was the theatre which inspired Fermi, after conversations with Bohr and Pauli, to propose a theory of beta decay.

SOLVAY FALLOUT

At the Solvay Conference, Lise Meitner presented new data on beta decay, which confirmed what she had seen before: the energy of the beta particle varied within a small range from one experiment to the next. This was the case for every example of a beta radioactive nucleus that she had measured. There could no longer be any doubt: either energy conservation does not apply, and quantum mechanics fails at the miniscule size of an atomic nucleus, or the beta particle is accompanied by an unseen neutral sibling, the total energy being shared between them in a random way.

Bohr had advocated the former, Pauli the latter. Meitner's news convinced Pauli that he was correct, and he duly promoted his idea of the neutrino at the Solvay Conference. Fermi debated with them how to resolve the issue.

Bohr agreed that quantum mechanics applied to electrons in atoms but had been sceptical about its relevance to the nucleus. Yet Gamow had used quantum mechanics to create a theory of nuclear alpha decay in which he had implicitly included Heisenberg's uncertainty principle. While this had initially just been a theory, the previous year, 1932, Cockcroft and Walton's experiments had demonstrated not only the truth of Einstein's equivalence of mass and energy, the key to determining which nuclei divest themselves of energy by radioactive decay, but also confirmed the validity of Gamow's quantum calculations. So, far from quantum mechanics not working at nuclear dimensions, the emerging evidence, discussed at Solvay, was that it does. By 1933, Bohr's option looked unlikely.

Fermi was ideally placed to resolve the impasse. Like Ernest Rutherford, Fermi had a gift for being able to visualise the atomic world. This enabled him to see a way through otherwise opaque problems, estimate the likelihood of various phenomena happening, and then work out the consequences. He had in this instance also the fortune to be in the right place at the right time: the right place, because of further news that arrived at the Solvay Conference from Frédéric Joliot-Curie, and the right time because during the previous years he had experienced several individual pieces of a jigsaw which only now merged to form a complete picture for him.

First of these experiences was the discovery of the neutron. Like everyone else, he had recognised the huge significance of Chadwick's discovery. Whereas most of the leading scientists had viewed the neutron as a composite of an electron and proton, Majorana had impressed on Fermi his vision of the neutron as a neutral proton so forcibly that

Fermi had incorporated this in his mental store of nuclear pieces. Majorana's visit to Germany in the first half of 1933 had convinced Heisenberg of this too, hence his paean to Majorana during his talk at Solvay. For Fermi, the nucleus was a cluster of protons and neutrons, twins, though not identical as they are distinguished by electric charge and slightly different masses.

Pauli's idea of an electron and neutrino had a pleasing symmetry to it, akin to that of the proton and neutron in Majorana's theory. Pauli had initially dubbed his creation a "neutron", but as that name had become attached to the nuclear particle, Fermi dubbed it "neutrino"—Italian for "little neutron"—aware that if Pauli was right the relatively small energy release in beta decay limited the mass of a neutrino to zero, or very near to it.

Fermi was attracted by Pauli's idea. Two further experiences helped clinch Fermi's vision. One of these was a further piece of experimental information that he had learned at the Solvay Conference; the other, which helped him construct his eventual theory, was inspired by his experiences years before with the British theoretical physicist Paul Dirac.

The key experimental news that helped fuse Fermi's insight came from Frédéric Joliot-Curie's talk. Although Irène had been attacked for her claims that the positrons and neutrons produced in their experiments were in some way evidence that the proton was a composite, Frédéric's presentation of their evidence for positron emission in radioactivity had been well received and had impressed Fermi. He realised its most profound implication to be that beta decay could occur in two distinct ways. Previously everyone had been exercised by the emission of negative rays which consisted of the well-known electrons, but now the Joliot-Curies showed that they had found examples where a positron emerged. The discovery of the positron had of course already created excitement prior to the conference, and now Frédéric was telling them that it could be produced naturally in radioactive decay. Furthermore, apart from the appearance of a positive positron instead of a negative electron, everything looked pretty much the same.

Fermi put these ideas together with the emerging picture of a nucleus made of protons and neutrons. He had a vision that the change from neutron to proton—or in reverse proton to neutron—could be the source of the beta particle of either negative or positive charge. It was now that Fermi had the key insight that neutrons, protons, and beta particles—whether electron or positron, it was just the electric charge that mattered—were the central players in these nuclear processes.

He knew that some force within the nucleus must act to convert a neutron into a proton, or vice versa. The nature of this new force was yet to be determined, but to construct a theory of how forces act in the world of quantum scales and relativity, he had a guide. In 1928 Paul Dirac had constructed a relativistic quantum theory of the electromagnetic field which describes the electromagnetic force. Now Fermi planned to use that as a basis for his theory of beta decay.

FERMI'S THEORY

Immediately after the Solvay Conference, Fermi set to work; by late December 1933 he was confident that he had found the answer. The first people to hear about it were Amaldi, Rasetti, and Segrè during a Christmas vacation weekend, skiing with Fermi and their families at Val Gardena in the Dolomites. Segrè recalls they had had a hard day on the slopes, which were icy. After many falls, he was heavily bruised and looking forward to a relaxing warm bath when he received an invitation to join Il Papa in his hotel room to hear him share "the best work he had ever done".[1]

Amaldi recalls how there was only one chair in the room. Fermi sat in the middle of his bed with crouched legs while the three of them sat around him on the edge of the bed, their necks twisted trying to see what he was writing on a piece of paper that he balanced on his knees.

He announced that he had solved the problem of nuclear beta decay. His starting point was to assume, like Majorana and Heisenberg, that all atomic nuclei consist of heavy particles: protons and neutrons.

The question then was how are the electron and neutrino created in the process of beta decay? Fermi explained that some previously unknown force must be at work and that he had built his theory of it by analogy with what Paul Dirac had done for the electromagnetic force in 1928.

Historically the electromagnetic force had been described by the equations of James Clerk Maxwell in 1865. Maxwell's theory included the famous prediction that electromagnetic radiation and light are one and the same. In Dirac's quantum theory of the electromagnetic field, radiation consists of photons which can be created, such as by the striking of a match, or can disappear into oblivion, as when they enter the retina of your eyes, their energy transforming into electrical impulses that stimulate vision. There is nowhere in Maxwell's theory where the emission and absorption of photons by atoms takes place, but in Dirac's formulation this is a key feature. For Dirac, an excited atom sheds energy by emitting a photon, which did not pre-exist in the atom but was created in the act of radiation. Fermi had basically taken Dirac's ideas and applied them to an excited nucleus, to spawn not photons but pairs of electrons and neutrinos, which too did not pre-exist in the nucleus. This was something different and completely new.

In essence, Fermi was proposing a new force of nature—known as the weak nuclear force—that only acts over distances smaller than an atomic nucleus.[2] Whether the force converts a neutron to a proton or vice versa is determined by the configuration of neutrons and protons within a nucleus; whichever process sheds energy to increase nuclear stability is the one that occurs.

Key to this whole construction was the adoption of Pauli's thesis that the electron is indeed accompanied by a neutral lightweight particle: the neutrino. Fermi explained that at the instant the change happens, the energy released materialises into a neutrino and an electron (or a positron). The total energy is fixed, the amount determined by the difference in binding energies of the initial and resulting nuclei. A fraction of the energy liberated in the beta decay escapes the experimentalists' apparatus if it is carried away by a lightweight electrically neutral neutrino. The way that energy is apportioned between the electron and the neutrino is predicted by the theory. Fermi calculated this for them and showed that if the mass of the neutrino is extremely small, the electron energy can take on a range of values, consistent with what Meitner and others had experimentally found. Fermi showed his three friends how to estimate the beta-decay lifetimes of several radioactive nuclei; these too agreed with experiment. There was no doubt: Il Papa had solved the mystery of beta radioactivity.

A century later, we know that he had correctly identified the actors and his ideas are in large part still valid. He wrote a paper in December 1933 entitled "Tentative theory of beta rays" and sent it to the leading English-language scientific journal *Nature*. Astonishingly, the editor rejected what is regarded as one of Fermi's greatest pieces of theoretical physics, having received advice that the manuscript contained "speculations too remote from reality to be of interest to the reader".[3] Half a century later, *Nature*'s board admitted this decision was possibly their greatest ever blunder. A preliminary article appeared in Italian in *La Ricerca Scientifica* and soon after his full paper was published in German in *Zeitschrift für Physik*, but never in English.[4]

Among the many physicists impressed by Fermi's theory were two young Germans, Hans Bethe and Rudolf Peierls. Peierls, born in 1907, was a bespectacled, slightly built man with a distinct overbite, while Bethe, a year older, had a high forehead from which hair rose straight upwards like newly sprung grass, all topping a square face with a broad mouth that carried a sense of permanent mischief. Like many leading theorists from central Europe, Bethe and Peierls had Jewish backgrounds and were even now becoming part of the growing diaspora escaping the tyranny of the Nazis. By 1934 they were in England, at Manchester University; Peierls eventually settled at Birmingham University, and Bethe at Cornell in the United States. Having fled fascism, they would later play central scientific roles in plotting its downfall, but in 1934 they were postdoctoral physicists already recognised as leaders of the younger generation applying quantum mechanics to atomic physics. They now made their entrée into the new field of nuclear physics.

Their opening salvo addressed the simplest composite nucleus where a single proton and a neutron link together to form the isotope of hydrogen known as a deuteron. Using quantum mechanics, they calculated the circumstances under which electromagnetic radiation in the form of photons would disintegrate a deuteron into its constituent neutron and proton, known as the *photo-effect*. The deuteron had been discovered in 1931, and following his discovery of the neutron, Chadwick was interested in knocking the neutron out of the deuteron. In 1933, during a visit to Manchester University, Chadwick had a chance meeting with Bethe and Peierls. He challenged them: "I bet you can't make a theory of the deuteron photo-effect".[5] Unknown to the pair, Chadwick had already done some experiments, and their theoretical predictions agreed with much of what Chadwick had found, further proving that quantum mechanics applies at nuclear dimensions.

Peierls had been present at the Solvay Conference and heard the talks by Meitner, Frédéric Joliot-Curie, and Heisenberg, along with the discussion about Pauli's postulated neutrino, which were the foundations of Fermi's theory. Enthused by Fermi's creation, Peierls and Bethe now examined its implications within quantum mechanics. They were impressed with its compelling simplicity. That a neutron could convert into a proton in conventional beta decay emitting a negatively charged electron was now matched by the Joliot-Curies' demonstration that positrons could be emitted too. In Fermi's theory this would arise from a proton converting to a neutron. Bethe and Peierls realised that if Fermi was correct, quantum mechanics implied that a similar conversion of the nucleus should arise from the absorption of a negatively charged electron. They used the equations of quantum mechanics to compute the likelihood and its character.

They also discussed the reverse process, in which a neutrino is absorbed by a nucleus and an electron is emitted. Using Fermi's theory, they calculated the chance of this happening, only to discover that it is extremely small, indeed so small that, as they pointed out, a neutrino could pass through the whole of the Earth with only a negligible chance of being absorbed on the way. Their paper suggested that the neutrino could be a particle that would forever remain elusive.

These conclusions made the concept of the neutrino on the borderline of whether it was indeed remote from physical reality, as the editor of *Nature* had suggested. However, contrary to the editor's opinion, Fermi's theory certainly contained speculations of great interest to the reader. Fermi regarded this as his greatest theoretical work, and today, thanks to advances in technology, the existence of the neutrino is not only established but the neutrino has become an essential tool in experimental physics.[6]

So, by 1934, nuclear physics had matured. That the nucleus is composed of neutrons and protons was established, and quantum theories explaining the origin and character of both alpha and beta radioactivity had been developed. This spontaneous energy release from the nucleus was now understood, at least theoretically. Whether and how it could be made to happen faster to provide a useful source of large-scale energy, however, remained the big unknown.

The arguments with the editor of *Nature* had exhausted Fermi to such an extent that he decided to switch from theory to experiments. His wife recorded he planned this to be "a pleasant diversion" for a short while.[7] In fact, this change of focus lasted for the rest of his life and would open the way to releasing nuclear energy.

12

Third Time Lucky

Irène and Frédéric Joliot-Curie left the Solvay Conference triply deflated. By the end of 1933 they had been exploring the uncharted inner space of the nucleus for four years. Having missed both the neutron and the positron, their consolation prize of positron emission accompanied by neutrons in radioactivity had now been publicly ridiculed. They were as sure that they had seen positrons emitted with neutrons as Lise Meitner was that this did not happen.

But less than three weeks after the conference, Irène received a letter from Meitner who had "looked again" using a more powerful source of alpha particles and "reached the conclusion that your interesting views on the disintegration of aluminium are correct".[1] Frédéric later recalled how this news gave him and Irène "two months of tranquillity" after the trauma of Solvay.[2] The tranquillity was terminated on 11 January

1934 when, stimulated by their colleague Francis Perrin, Frédéric and Irène made their momentous discovery of artificial radioactivity.

During December the Joliot-Curies had investigated positrons by use of a cloud chamber. The positrons left short, wispy trails as they emerged from the nuclear source and were almost immediately annihilated when they encountered the ubiquitous electrons in the chamber's vapour. Frédéric was able to photograph the trails, which provided an image of an event at a given instant. To record the evolution of the process over a period of time, he turned to a Geiger counter. This new focus suggests that Frédéric and Irène "had finally embraced Perrin's interpretation" of their results.[3]

In December they learned that Perrin had published a paper about Pauli's neutrino hypothesis, and early in January came news of Fermi's theory of beta decay. Suddenly there was a theoretical framework for understanding beta decay and the appearance of positrons. This stimulated Frédéric and Irène to resume their earlier experiments, ending the "two months of tranquillity".

Meitner's letter gave them a hint that the alpha particles used to bombard the target nuclei in her experiment had a different energy to those in the French version. They decided to conduct a series of experiments in which they varied the particles' energy.

The Joliot-Curies achieved this by varying the distance between the target and the source of alpha particles. The more air the alphas had to fight through, the slower they would be upon reaching the target. They had guessed right, for when the source was near to the target, the alphas blasted it with the highest energy and ejected neutrons and positrons; when they moved it further away, the alphas' impact was reduced, and the emission of neutrons ceased. This resolved their earlier standoff with Meitner.

The saga of nuclear energy began with serendipity, for both Wilhelm Röntgen and Henri Becquerel, but the real fortune was that chance made the revelation to prepared minds. The ability to recognise that some seemingly inconsequential happening should be followed up and not ignored can separate winners from also-rans. So was now the case for Frédéric Joliot-Curie. Had the loss of face arising from Meitner's intervention not happened, or Meitner's follow-up letter not been sent, he might never have made the refinement of varying the alpha particles' energies. And had he been less vigilant, he might have missed the diamond that now emerged from the gloom.

Frédéric was an outstanding experimentalist, alert to the unexpected. When the source was moved farther away, neutron emission ceased but the Geiger counter continued to emit crackles: something was still producing positrons but without neutrons. He repeated the experiment, but this time, after suitably irradiating the target and detecting both neutrons and positrons together, he removed the alpha source entirely. The same thing happened: neutrons ceased but the counter continued to click.

He realised that there were two possibilities. The exciting one was that positrons were still being produced without any neutron accompaniment. The second, more mundane, was that the Geiger counter's erratic performance showed it to be faulty.

Frédéric made this key discovery on the afternoon of 11 January 1934, but as he and Irène had an unavoidable social engagement that evening, he had to leave the laboratory with this key issue unresolved. Fortunately, there was a young German named Wolfgang Gentner working at the Radium Institute that year who was an expert in Geiger counters. Frédéric asked him to check the laboratory's counters while he and Irène were out that evening to see if the devices were working satisfactorily. The following morning the Joliot-Curies arrived at the laboratory to find a handwritten note from Gentner confirming that the Geiger counters were indeed in perfect working order.

So, the first possibility was the answer: the alpha particles had kicked out neutrons and positrons, but when the source was removed, only the neutrons ceased. The positrons continued to trigger the Geiger counter, its crackle dying away gradually to half the intensity in about three minutes.

If there was any doubt remaining, it was dispelled when he found similar results with targets of boron and magnesium. Once again, removal of the source left a signal of positrons dying away like before, the key difference being that in the case of magnesium it died to half intensity in a little over two minutes, and in the case of boron more like a quarter of an hour. Everything began to look as if somehow radioactivity had been induced in the target and that Frédéric was observing the half-life of a synthesised radioactive nucleus in much the same way as beta radiation is emitted from a naturally radioactive element. In the Joliot-Curies' case the beta decay involved the positively charged positron rather than the negatively charged electron.

This appeared very similar to what Francis Perrin had suggested at the Solvay Conference on theoretical grounds might be possible when alpha particles impact atoms of aluminium. Perrin's thesis could also explain the Joliot-Curies' results with boron and magnesium. If that interpretation was correct, then in the case of boron (five protons and five neutrons) the alpha particle (two protons and two neutrons) impact would have created a neutron and nitrogen-13 (seven protons and six neutrons). The normal, stable form of nitrogen (nitrogen-14) contains seven protons and seven neutrons. Nitrogen-13 is unstable. If it were to decay by positron emission, as seemed the case in their experiment, this would leave the natural stable isotope of carbon: carbon-13 (six protons and 7 neutrons)—see Figure 7 (top).

The next step would be to prove these transformations were indeed the explanation. Years of experimenting with alpha particle beams showed that only about one in ten million of them would hit and successfully cause a transmutation. Even with the Joliot-Curies' superb polonium source, which produced copious numbers of alpha particles, this meant that no more than about one hundred thousand atoms of the radioactive elements were being formed. In terms of mass this amounted to less than a millionth of a billionth of a gramme, too little to detect by chemistry alone. The trick would be to follow the radioactivity as the new radioactive element joined with conventional atoms of the same element in some chemical compound.

In the case of boron, where the Joliot-Curies suspected a radioactive isotope of nitrogen resulted, they chose as their target boron

$$\alpha_4^2 + B_{10}^5 \rightarrow n_1^0 + N_{13}^7$$

$$\hookrightarrow e^{+(1)}_{(0)} + C_{13}^6$$

$$\alpha_4^2 + Al_{27}^{13} \rightarrow n_1^0 + P_{30}^{15}$$

$$\hookrightarrow e^{+(1)}_{(0)} + Si_{30}^{14}$$

Figure 7. Inducing radioactivity in boron (top) and aluminium (bottom). Superscripts denote electric charge and subscripts the total number of neutrons and protons.

nitride, a molecule consisting of one boron atom and one nitrogen atom. If they were correct, alpha bombardment should convert up to a hundred thousand atoms of boron into radioactive nitrogen. With a half-life of fourteen minutes this left enough time for them to heat the irradiated boron nitride with caustic soda. A chemical reaction then produces gaseous ammonia, NH_3, whose nitrogen atoms do not care whether they are conventional or the radioactive form. In this case the gaseous ammonia would carry away the radioactivity, leaving behind the boron compound with no nitrogen and no radioactivity. Indeed, the Geiger counter remained quiet, proving that the radioactivity had left the boron. Meanwhile the ammonia gas was diverted by a tube and collected, taken across the lab away from any residual radioactivity, and tested. When the Geiger counter was pointed at the ammonia, it burst into life proving that some of its nitrogen atoms were radioactive. Here was a direct proof that their irradiation of boron had created a novel radioactive form of nitrogen.

The Joliot-Curies took a similar approach in the case of aluminium (Figure 7, bottom). Here they dissolved the aluminium foil in hydrochloric acid, which produces hydrogen gas. The radioactive phosphorus fused with the hydrogen to form phosphine (three hydrogen atoms for every one of phosphorus, PH$_3$). Like the ammonia (NH$_3$), this was collected in a tube which as before made the Geiger counter crackle, proving that the radioactivity had indeed gone with the phosphorus and departed from the original aluminium sample.

This was revolutionary work.[4] Radioactivity had for more than three decades been an intriguing but rare phenomenon, exhibited by but a handful of elements. Frédéric and Irène had now discovered that it is possible to alter the nucleus and induce radioactivity in otherwise inert materials such as mundane pieces of aluminium foil. Had natural radioactivity or use of expensive atom smashers been the only possible ways for atomic nuclei to liberate energy, Rutherford's allusion to "moonshine" would have been correct. Rutherford and the Cambridge scientists had established that the atomic nucleus can be manipulated into novel forms by brute force, but the French pair had done much more: using rays emitted spontaneously by a thimbleful of polonium, they had taken the first step to releasing some of the vast energy that had been locked within quiescent nuclei for billions of years. Their experiment showed that it is possible to liberate part of an atom's latent nuclear energy, potentially in amounts far exceeding anything known to chemistry. Here was the first proof that nature's chronometer, where emission of nuclear energy is sluggish at best, can be altered. Emilio Segrè later described this innovation as one of the most important discoveries of the twentieth century.

The Joliot-Curies had revealed a vista with a wealth of opportunities for medicine, science, and technology. They received the Nobel Prize in chemistry for their discovery in 1935. At last, they had reached parity with James Chadwick, whose discovery of the neutron won him the physics prize that year. Upon receiving the award, Frédéric presciently remarked that by modifying atoms this way it might be possible to "bring about transmutations of an explosive type". As he elaborated: "If such transmutations do succeed in spreading in matter, the enormous liberation of useful energy can be imagined."[5]

13

To Uranium and Beyond

"WE WILL ALL HAVE TO LEARN ITALIAN"

When Enrico Fermi heard about the Joliot-Curies' discovery, he decided to try to induce radioactivity for himself. Following the trauma of his rejected paper on the theory of beta decay, the experimental work reinvigorated him.

Like the French, Enrico Fermi had no access to atom smashers. Unlike the French, however, he had Ettore Majorana. In hindsight Majorana's idea to use neutrons was obvious, but at the time it was radical. As he pointed out, the nuclear electric shield that repels positively charged invaders like alpha particles or protons is impotent against neutrons, which are electrically neutral. The implication: neutrons can enter the nuclei of all elements without penalty. Instead of alpha particles, use neutrons. Majorana said it; Fermi worked out how to do it.

The problem is that free neutrons are very rare. To create beams of neutrons you first must bombard atoms of beryllium with alpha particles; that is the way the neutron was discovered after all. Because most of the alpha particles miss the beryllium nuclei, the process produces only one neutron for about every ten thousand alphas. This seemed so wasteful that Majorana and Fermi could easily have dismissed the project. But as they realised, neutrons could enter the nuclei of even the heaviest atomic elements. The fact that the French duo were able to achieve anything at all was primarily because they had such powerful sources that some of the alpha particles got through the electrical resistance. Even so they were only able to reach as far as aluminium, atomic number 13; the resistance of silicon's fourteen positive charges was already too much. Even with the most powerful alpha sources, the nuclei of elements beyond aluminium would be forever out of reach.

Neutrons are expensive to make, but once you have them, they are ideal. Stable isotopes are those where the number of neutrons is optimised, so when a neutron hits, the product is often an unusual isotope of the element with an abnormal number of neutrons accompanying the protons. Nature craves stability, and these unstable isotopes adjust to make more stable forms by emitting radiation.

Fermi designed the experiment with artistic creativity. His neutron source consisted of a glass tube containing radon and beryllium, the radon providing alpha particles which hit the beryllium and ejected neutrons. Because these neutrons sprayed in all directions, he engineered his samples so that they surrounded the source entirely. He did so by using a hollow cylinder, which he coated with the element to be irradiated, and then placed the source inside. He let the neutrons do their work for a few minutes or a few hours, depending on the needs of the experiment, and then removed the source, inserting in its place a Geiger counter to measure any induced radioactivity.

The setup was intrinsically simple, but Fermi realised there was a catch: the neutron source was itself highly radioactive, much more so than the amount of radioactivity that would be induced in the samples. So to ensure that detection of radioactivity was genuinely from the sample and not from the source itself, the measurements were done in a room at the far end of a long corridor, well away from the experiment.

On 20 March 1934, Fermi accomplished his goal, inducing radioactivity in aluminium by means of neutrons, before doing the same with fluorine. In each case the balance of neutrons and protons in the initially stable target atoms is delicate, and the invader disturbed it. The new grouping gave up energy and attained equilibrium by readjusting the ratio of neutrons and protons, which is achieved by emitting an electron or positron. Fermi announced his discovery in a letter to *La Ricerca Scientifica* on 25 March titled "Radioactivity induced by neutron bombardment".

Next, he attacked heavier elements. This would require a team effort, so Fermi co-opted Amaldi, Segrè, and Rasetti. As the act of irradiation and decay also changes an element's chemical identity, Fermi needed a trained chemist to separate these radioactive nuclei from the mix. Fermi enrolled chemist Oscar D'Agostino into the team for that very purpose.

The radioactive element's half-life was in some cases very short, so there was no time to lose in getting the newly activated sample to the distant room containing the Geiger counter. Immediately after the irradiation of a sample, Fermi—always competitive and eager to take part in a race—and one of his colleagues, sporting dirty grey overalls, would rush down the corridor with the precious hot metal clutched to their chests. This proximity to radiation may well have caused the stomach cancer from which Fermi died and others became ill.

Within two months of the Joliot-Curies' groundbreaking paper, Fermi's team had irradiated elements deep into the periodic table as far as barium, atomic number 56, which is already four times further than the French duo's limit of aluminium at number 13.[1] The upper reaches of the table, all the way to uranium at 92, now beckoned. By the summer of 1934 Fermi's team had irradiated sixty elements, inducing radioactivity in about forty of them.

Fermi sent news of their achievement to laboratories in other countries. Its importance was immediately recognised. Rutherford, the experimentalists' experimentalist who was always sceptical of theorists, jokingly replied: "Congratulations on your escape from theory". In the United States, Isidor Rabi, the young star experimentalist at Columbia University in New York, whose first language had been Yiddish and then English, ironically remarked: "Now we will all have to learn Italian".[2]

Fermi's team produced a series of papers describing the work and their results. All five members of the team were included as authors of the papers, which was a first for physics and perhaps of science in general. Previously throughout history one or at most two people would combine in a research project; here for the first time was teamwork, a new way of doing science in increasingly large collaborations. In addition to the fundamental physical discoveries coming from his laboratory, Fermi was also changing the way that research and development in nuclear physics and technology would be done in the future.

EXTRATERRESTRIAL

The heaviest element found on Earth is uranium, atomic number 92. By 1934 it was clear that the nuclei of all elements are made of neutrons and protons and that the relative number of those two constituents determines the stability of the resulting isotope. The rules determining what makes one isotope more stable than another and the mechanisms whereby unstable isotopes attain stability through alpha or beta decay were yet to be determined. Nonetheless, there was no obvious reason why more than ninety-two protons could not clump together if enough neutrons were also present to help stabilise the system. The absence of such material on Earth was assumed to be because such combinations are highly unstable, such that any present at the Earth's formation had died out by radioactive decay, eventually reaching the relative stability of uranium.

Uranium's half-life of 4.5 billion years is the reason why so much of that element has survived in the Earth's crust, but even here over aeons radioactive decay has eroded uranium and its progeny, eventually leading to the ultimate stability of lead. Fermi's hope was to reverse history, in effect, by irradiating uranium with neutrons and momentarily producing long-lost *transuranic* elements, atomic numbers 93 and above.

The results were unusual, the products of the reaction having a wide range of half-lives and responding to chemical tests in ways that suggested something was being overlooked. Fermi insisted that D'Agostino compare the results against everything in the periodic table from uranium down to lead, the "ground level" of stability underpinning the sequential radioactive decays of uranium. Finding no matches all the way to ground level, he believed that the source must be higher up: new transuranic elements.

Fermi was initially careful not to claim more than he could prove, but his patron, Senator Corbino, grew overexcited. In a paean to the glories of fascism, Corbino cited this work as the first example of a renaissance in Italian science with the creation of elements never seen on Earth. The world assumed Fermi sanctioned this enthusiastic announcement. Nuclear physicists everywhere set out to repeat what he had done.

In mid-1935, Marie Curie died, cancer ridden from years of exposure to radiation. Irène was also beginning to suffer ill health for the same reason. Following her mother's death, Irène and Frédéric, with the added prestige of becoming Nobel laureates, took over control of the Radium Institute. Frédéric was made a professor at the Collège de France and put his energies into developing a cyclotron, following Ernest Lawrence's strategy of a brute-force attack on the atomic nucleus. Irène still smarted from Lise Meitner's remarks at Solvay, notwithstanding her subsequent successes with artificial radioactivity.

She decided to pursue one of the active by-products that Fermi had reported.

She invited a young Yugoslavian physicist, Pavle Savić (sometimes written as Paul Savitch), to assist her. They successfully isolated the substance and established its half-life to be three and a half hours. This was long enough to analyse its chemistry, but when they did so, the results puzzled them: the substance appeared to behave like lanthanum, element number 57, almost halfway back down the periodic table! The idea that an element in the middle of the table could suddenly appear when they were irradiating uranium at number 92 made no sense within the existing understanding of nuclear physics and radioactivity. This physics prejudice led them to ignore the chemical evidence.

Trapped in this false belief, they repeated the experiment many times, and while they showed conclusively what the substance was *not*, they were never able to identify with certainty what it actually was. Even so, the chemistry gave enough clues to make Irène certain that it was not transuranic. The best they could conclude was that it appeared to be similar to thorium, element number 90. Their announcement that it "might be a form of thorium" revealed their uncertainty.[3]

In Berlin, Otto Hahn and Lise Meitner too faced the conundrum of the mess of radioactivity that Fermi took to be evidence of transuranic elements. Hahn's radiochemical expertise enabled him to confirm that the radioactive substances could not be from any element below uranium in the periodic table, all the way down to mercury at number 80. He wrote a letter to Irène saying that he and Meitner had isolated the three-and-a-half-hour substance and proved conclusively it was not thorium. However, as he admitted, he could not say what the substance was. Hahn drew the same conclusion as Fermi: they had created transuranic matter.

Lise Meitner was less convinced, although she did not adopt a strong negative position. However, the German chemist Ida Noddack did. She wrote in a German chemical magazine that under neutron bombardment heavy nuclei like uranium might "break into several large fragments" which are isotopes of known elements and not neighbours of the irradiated ones. As Noddack pointed out, no one had checked whether elements *lighter* than mercury might be responsible. Although she did not say so explicitly, she was open to the possibility that uranium nuclei had split in two, what is today known as "fission".[4]

Her remarks were regarded as so scientifically naive that she was not only criticised but personally vilified. She sent Fermi a copy of her paper, but he seems to have shrugged off her insight and continued his pursuit of transuranic elements. Otto Hahn reacted more strongly, describing Ida as an awful person, ridiculing her on scientific grounds and portraying her as an industrial chemist who was naive about nuclear physics. It was by then common knowledge that radioactivity involved but a gradual move, one or two places along the periodic table. Even a sequence of alpha or beta decays would only reach the stability of lead, at atomic number 82, before ending, which is why Fermi argued that nothing could incrementally go beyond lead. Hahn and Meitner had repeated his experiment, verified his results, and drawn similar conclusions, and as a small extension had checked two places beyond lead, to mercury at atomic number 80. Noddack's idea that uranium could be changed so radically as to produce elements lighter than this was unthinkable, indeed so unthinkable that Irène Curie and Savić had rejected their own evidence of lanthanum.

THE SCEPTICAL IDA NODDACK

Ida Noddack was not alone in her criticisms, though none of her male counterparts appears to have received the bullying negativity that she experienced. Why was she ignored, even abused? A standard answer to this question is that she was a woman. Certainly, as a married woman she was not well placed in Germany at that time where, following the Wall Street crash and the onset of economic depression, in 1932 the government passed a law to protect jobs for men: spouses had to quit their careers and become housewives. So it was that in 1934, Ida

Noddack, relegated to the status of *hausfrau* but continuing chemical research as an "unpaid collaborator" with her chemist husband Walter, criticised the great Enrico Fermi, Otto Hahn, and by implication the wider scientific community.

In the circumstances it is perhaps of little surprise that she was mocked for her perceived impudence, though the venom hints that misogyny was not the full explanation. While it is true that in the 1930s women were second-class citizens, their professional role subjugated relative to that of their male colleagues, the field of nuclear physics at that time already included some heavy-hitting women, most notably Marie Curie whose reputation was earned independent of that of her late husband and whose international stature outstripped his. Their scientist daughter Irène gave credibility to her husband Frédéric Joliot-Curie, and not least there was Lise Meitner. Indeed, Otto Hahn, whose criticisms of Noddack had been vociferous and uncharacteristically aggressive, had worked with Meitner for two decades and they were recognised as equals in the scientific community. To dismiss the physicists' reactions as simply misogyny is too simplistic. Deeper forces were at work.

First, the opprobrium contained elements of a professional sense of ownership. Noddack was not a nuclear physicist but a chemist; while she was an expert on the periodic table and chemical properties, she had no track record in radiochemistry let alone the intricacies of nuclear physics. Second, her work in chemistry where she had co-discovered two elements was so controversial that Emilio Segrè described it as "plain dishonest".[5]

Born Ida Tacke, in the Northern Rhineland in 1896, she had become interested in science at school. Having no wish to become a teacher, and as there were few jobs in physics, upon graduating in chemical engineering at Berlin's Hochschule für Technik in 1918, Ida became an industrial chemist at the AEG turbine factory. Meanwhile her future husband, Walter Noddack, was a brilliant scientist in the University of Berlin's chemistry department. In 1922 he became director of

Germany's Imperial Physical Technical Bureau. In their chemical division he was looking for missing elements in Mendeleev's periodic table. It was in 1924 that Ida resigned her post at AEG to become an unpaid collaborator of Walter at the Technical Bureau.

Walter was focused on two missing elements, atomic numbers 43 and 75, which are in the same column of the periodic table as manganese at number 25. Manganese is found in columbite—a brownish black mineral—and so the two of them examined samples of the ore. Using X-ray equipment of Berlin physicist Otto Berg, in 1925 they announced the discovery of the two missing elements in the ore. Number 75 they named rhenium in honour of Ida's origins on the Rhine and published this in a paper written jointly with Berg. A separate paper published soon after by Ida and Walter alone announced the discovery of element 43, which they named masurium after Walter's birthplace. This is when their troubles began.

Those discoveries were soon disputed when others were unable to reproduce their results. The absence of Berg's name on the second paper also raised eyebrows as to the solidity of their evidence for masurium. Reproducibility is key to the advance of science, an inability to do so being a first step towards negation of a claim. Walter and Ida surveyed eighteen hundred ores in the hope of getting weighable quantities. They succeeded in accumulating a gramme of rhenium but no masurium. While undisputed as the discoverers along with Berg of rhenium, their claims for element 43 remained unresolved.

They married the following year and continued to work together. Some gossip among the male scientific community however hinted that she and Walter were more bedfellows than a real scientific team of equals. Contrasting this misogynistic attitude were positives: they obtained a patent for rhenium as a coating for lamp filaments and were nominated jointly for the Nobel Prize in chemistry on three occasions: 1933, 1935, and 1937.[6]

Meanwhile, some in the physics community who were knowledgeable about the masurium claim were sceptical, describing the work as "weak". The only record of explicit concern was an exchange with Emilio Segrè later, in 1937. Segrè had by then undisputedly identified element 43—now named technetium—in metals that had been irradiated with protons at Ernest Lawrence's accelerator in California. Segrè met Walter Noddack and attempted to compare their results. Segrè asked if his result agreed with Noddack's earlier claims about masurium: "Yes" came the reply. Segrè's query, "Have you found more about its chemistry than we have?" elicited the monosyllabic answer "no".

"How much masurium do you have?" he asked, to receive the unsatisfying response that the Noddacks had sent their sample to Cambridge University for isotopic analysis. This surprised Segrè, and the fact that Noddack could provide no information on what had become of the samples made him increasingly suspicious. He asked to see the X-ray plates with element 43's characteristic spectrum, only to be told they were broken and so they were unavailable. Segrè asked Noddack why they hadn't made more plates and recalled, "I could not obtain a clear answer".[7]

Segrè decided the Noddacks were either deluded or were just having the conversation in the hope that further results from Segrè would clear up any doubts that they had. Whatever the real agenda behind the conversation, Segrè was left with a low opinion of the Noddacks' integrity, leading to his subsequent one-line summary: "plain dishonest".

Segrè's opinions were formed some years after the events of 1934, but they seem like echoes of some of the impressions in the German physics community at that time. These appear to be based less upon questionable chemistry than upon Ida Noddack's perceived rejection of the advances that had taken place in nuclear physics. By 1934 the field had provided chemistry with tools to break up elements and make new ones. Chemical analysis alone was deemed passé. The Noddacks' approaches were regarded by the nuclear physicists as shortsighted, which helped discredit them in that community.

None of this fully explains the apparent venom that ensued from Ida's questioning of Fermi, however. The anger was driven by remarks that she published earlier that year during celebrations of the centenary of Mendeleev's birth. First came a lecture that she had given about gaps in the periodic table. She stubbornly clung to the old definition of elements based upon atomic mass rather than atomic number, even though for nearly two decades atomic number had been recognised as the correct way to organise the atomic elements. And in her talk about gaps in the periodic table, there was none at number 43: the Noddacks persisted with their claim to have discovered masurium. Worse though was an article in which she developed these ideas. Here she argued for a new system of classification based upon isotopes which, she claimed, revealed gaps between the ninety-two chemical elements and which, when filled, would increase the total to about 280. Had she stopped there she and her theory, based on analogy with the behaviour of electrons in atoms, might have been gently ignored, but she overplayed her hand, dismissing the current interpretations of nuclear physics as "dogma" which will "follow the fate of all dogmas and disappear".[8] To say that she was regarded as a maverick would be an understatement.

Ida Noddack's criticism was not that of a naive outsider entering the world of nuclear physics but was based in her solid grounding in chemistry. She knew that the element beyond uranium, now known as neptunium, would lie in a column of the periodic table one place below rhenium. As discoverer of rhenium, she knew everything about its behaviour, from which she was able to anticipate what the chemical properties of neptunium or other potential transuranic elements in the vicinity would be. None of what Fermi was finding in his experiment looked anything like them. As a chemist she had identified a loophole in his investigation: there were elements that he had not checked.

We now know that her idea of nuclear fission was correct, and ironically it was her expertise in element hunting that had led to her criticism. Nonetheless, her scepticism was dismissed by the physicists as hogwash. By default, scientists and, thanks to Corbino, the world concluded that Fermi had produced transuranic elements.

SLOW NEUTRONS

Fermi's team had induced radioactivity in nearly half of the elements that they had examined, some showing more intense activity than others. The next step would be to quantify the phenomenon. As the irradiation of silver produced radioactivity roughly in the middle of the range that they had measured, Fermi proposed that they quantify the relative amounts for the various elements and compare the results to that of silver. This seemed a good project for a new member of the group, Bruno Pontecorvo, a young, suave graduate from Pisa who had recently joined the team. Fermi gave him the task, supervised by Edoardo Amaldi.

They repeated the experiments that the team had done before, but this time rather than just recording the amounts, they also measured the intensity. The problem was that Pontecorvo was unable to get the same results for a particular element from one measurement to the next. He discussed this with Amaldi who repeated the experiment and found the same anomaly.

The two youngsters told Franco Rasetti who despaired at their lack of competence until he went through the procedure and found the same! This was clearly no mistake but a hint of something utterly unexpected. Pontecorvo later recalled that there were wooden tables in the laboratory which had miraculous properties: "Silver irradiated on these tables became much more radioactive than when an identical sample was irradiated on the marble table tops in the room."[9] The problem was to understand why; Fermi was informed.

Much of this saga has involved incremental discoveries, like building a tower one layer at a time. Others have involved chance or singular breakthroughs, such as Becquerel's discovery that started it all or Rutherford fixating on his "capricious variation". As Rutherford had pursued his anomaly, so Fermi now did this one.

While Fermi was preparing the experiment, he noticed a piece of paraffin wax lying around. He recalled that without any conscious reason he decided to use this and place it between the source and the silver. He discovered that the radioactivity of the silver was then much higher than it had been without the paraffin. It was the morning of Saturday 20 October 1934. Fermi showed his results to Amaldi, Rasetti, and Pontecorvo, but then it was time for lunch. During the lunchbreak Fermi continued to ruminate.

He visualised a neutron in flight, bumping into atoms in its surroundings and slowing down. His uncanny ability to imagine a neutron's view of the surroundings then made everything clear to him. Hydrogen is present in water, which is found in wood but not marble. It is also present in paraffin. A lightweight atom such as hydrogen would be especially good at reducing the neutron's speed—the impact of a neutron on a proton of the same mass would cause the original momentum to be shared between them roughly equally. Slowing the neutrons was key.

Whereas positively charged alpha particles need high speeds to penetrate the repulsive electric fields that surround the nucleus, neutrons don't need any such aid. For neutral neutrons, impervious to electrical impediment, the rule is: the slower the better. Lumbering neutrons, known as *thermal* neutrons because they have slowed to the point that their motion is no more than thermal agitation, remain in the vicinity of the target atoms for longer than fast-moving ones, giving them a greater chance of being captured and activating the sample. Fermi had experienced an epiphany: slow neutrons are especially good at inducing nuclear reactions.

This was a remarkably bold conclusion. Up to that time, the received wisdom had been that the harder you hit a nucleus, the more likely it is to fragment. Fermi now realised this wisdom was wrong. Nature is more subtle; radioactivity would become especially strong if there were some means of slowing the neutrons radically.

The way to test Fermi's hypothesis would be to irradiate the samples underwater. There is a delightful story of how they playfully submerged samples in Senator Corbino's goldfish pond, which was in the courtyard adjacent to the laboratory. This was told by Fermi's wife Laura many years later, though if true at all the tale is somewhat embellished.[10] The more mundane reality was that a cleaner had left a bucket of water in the laboratory; Fermi immersed some samples of caesium and rubidium nitrate in the bucket and irradiated them overnight. The next day they showed high amounts of radioactivity. Convinced he was on the right track, Fermi irradiated submerged samples of five other elements during the next night. These too became highly radioactive.

It is likely the team made a demonstration for Corbino, and did so by immersing silver in his pond, from which Laura Fermi's story evolved. Corbino was very excited by their results and was present when they were written up later in the evening of 22 October. This was done at Amaldi's home and was an animated occasion.

The excitement at the discovery was palpable, and there was much debate about exactly what to say in the paper. Corbino, a man of the world, had realised what the young scientists had not: their discovery could have industrial applications. Previously, the amount of radioactive material created using alpha particles or neutrons had been trifling; now, the slow neutron technique could produce it a hundred times more abundantly, and the practical implications were tantalising. "Take out a patent" Corbino shouted, which they did.

Amaldi's son, Ugo, was then just a baby. Decades later he became a scientist himself and described to me how he was asleep upstairs on that fateful night. He recalled being told at several family gatherings that the next day his nanny had asked his mother whether the "Signori the night before had been tipsy". The occasion had clearly been very animated. In celebration, an iconic photograph of the group was taken (Figure 8). It shows, from left to right, D'Agostino, Segrè, Amaldi, Rasetti, and Fermi; the one person missing being Pontecorvo who, as the youngest member of the group, was on the other side of the camera having been assigned the task of taking the photograph.

They were yet unaware that the most singular implication of the slow neutron phenomenon would be the explosive release of energy in the atomic bomb. That would require two further breakthroughs. Ironically, one of these—fission of uranium—had already been achieved as Ida Noddack suspected, but the world at large had failed to realise.

Figure 8. From left to right: Oscar D'Agostino, Emilio Segrè, Edoardo Amaldi, Franco Rasetti, and Enrico Fermi. The photograph was taken by Bruno Pontecorvo. (Courtesy of Gil Pontecorvo and the Department of Physics, Sapienza University of Rome.)

14
Majorana's Vision

When Ettore Majorana returned to Rome in 1933 from his visit to Werner Heisenberg in Leipzig, he probably understood more nuclear physics theory than anyone in the world. Working with Gian-Carlo Wick, the other theoretical physicist in the Via Panisperna team, he now helped to develop a mathematical model to explain the stability of heavy nuclei. Essentially, the question he was attempting to answer was: which numbers of neutrons and protons give the most-stable configurations?

It takes but a moment to realise that there must be some guiding principle because the resulting number of mixtures of neutrons and protons found in stable nuclei are far from random. For example, although varieties of atomic nuclei differ merely by these numbers, the abundances of the resulting elements vary enormously. Oxygen, calcium, and iron are very common, while there are others such as rhodium, ruthenium, and holmium that you may never have heard of. The ones you think of first tend to be among the most common, while those unknown to you are among the rarest. While the names oxygen and carbon are in everyone's lexicon—part of our DNA even—astatine and francium amount to less than a gramme of the Earth's crust.

Although the names of the common elements are well known, the abundances of their isotopes are far from general experience. All isotopes consist of protons and neutrons, differing merely in the relative number of those constituents, but their numerical distributions are not random. By the 1930s it was already becoming apparent that there is a preference for isotopes to contain an even number of protons or of neutrons, rather than an odd number. Today this tendency is obvious.

First, there are more nuclei with an even total number of constituents than odd—some 150 versus 100. What's more, an even number can be built from either two evens or two odds, and of that 150 all but five are made from both an even number of protons and an even number of neutrons. Why should nature so dislike odd numbers that there are but five odd-odd nuclei as opposed to 145 even-even ones?

Majorana had implicitly answered this already with his explanation of the stability of the alpha particle where two protons and two neutrons in a magnetic grip lower their overall energy. This is a general truth: nuclei dislike singletons. Any lone proton or lone neutron seeks a partner to pair with, forming a more stable relationship. In the extreme example of an odd-odd nucleus, all nucleons will be paired but for one neutron and one proton. In such a case beta decay—of the lone neutron or proton—transforms to two of one variety and none of the other; in other words, beta decay stabilises the bundle by converting it to an even-even state.

So much for odds and evens, but what determines the relative abundance of neutrons and protons overall? The basic physical principles are easily seen. For example, suppose you have a collection of 240 nucleons in total. What relative proportion of neutrons and protons will nature adopt? You might guess that it will choose them equally, giving 120 of each. This might have been the case had neutrons and protons been more like identical twins, powerfully attracted when they touch, but protons are electrically charged and mutually repel one another. This adds energy to the mix, which creates instability and mitigates against an excess of protons. The element becomes more stable when one of the protons undergoes beta decay, converting to a neutron by emitting a positron.

Uranium is almost unstable. Its nucleus is large with about six nucleons across any diameter. This means that neutrons or protons on opposite sides of the nucleus are too far apart to mutually attract, whereas any protons push one another apart electrically wherever they are. So, a big, heavy nucleus needs an excess of neutrons to bind it. Smaller nuclei have less of a problem. The girth of barium or krypton, for example, is only three or four nucleons. The strong localised attraction is thereby more efficient, and relatively fewer neutrons are required to combat the protons' electrical disruption.

Empirically, no more than ninety-two protons can survive in a tight cluster, which is why uranium—element number 92 in the periodic table—is the largest stable element found on Earth. So, in the most abundant isotope, uranium-238, the ninety-two protons are clustered with 146 neutrons. Neutrons, which have no electric penalty, give an extra contribution to the mutual attraction among the members of the pack.

You must not add too many neutrons, however. A neutron has a slightly larger mass than a proton and so, thanks to $E = mc^2$, contains slightly more energy. Adding too many neutrons will add this unwanted extra energy to the nucleus, encouraging instability once again. In this case, beta decay of a neutron increases stability as it converts to a proton by emitting an electron. Majorana and Wick worked out a formula for the optimum combinations which make the most-stable nuclei. In typical Majorana style this work was regarded as obvious and not published.

The theoretical understanding of nuclear stability moved apace. In 1935 the German theorist Carl von Weizsäcker developed along similar lines a more refined semi-empirical mass formula, by which the binding energies of nuclei could be estimated and the likely fates of unstable nuclei determined. Those nuclei with an excess of protons shed an electrical charge by emitting positrons, converting one or more protons to neutrons. This explains what the Joliot-Curies had found. Likewise, a nucleus with too many neutrons could convert a neutron into a proton by emitting an electron. Nuclei with very large numbers of protons and corresponding huge electrostatic energy need an excessive number of neutrons to provide the attractive glue. Even so, by the time ninety-two protons have been gathered, as in uranium, the limit of stability has been reached.

However, there remained that elusive possibility of highly unstable elements beyond uranium.[1]

THE MYSTERY OF MAJORANA

Ettore Majorana disappeared in March 1938. He was presumed to have committed suicide and died by drowning after leaping from the Palermo to Naples ferry. No corpse was ever found, however, and his real fate remains a mystery to this day.

If indeed Majorana was dead at the age of thirty-one, he joins the likes of Roger Cotes in mathematics and Frank Ramsey in economics, each a genius rated as such by a true master.

Cotes, who died in the seventeenth century aged thirty-three, had applied Isaac Newton's ideas to the orbits of comets and made a theory of equinoxes, which formed the core of Newton's second edition of his *Principia*. When Cotes died, Newton said, "If he had lived, we would have known something".[2] Ramsey was a student of mathematics in the 1920s who criticised aspects of John Maynard Keynes's economic theory and died aged twenty-six. Keynes rated Ramsey as being very much more clever than he was, in accord with the description "there was something of the Newton about him".[3] So Majorana, rated by Fermi as one of the magicians alongside Isaac Newton, is a twentieth-century physics analogue of these supernovae whose brilliant light was extinguished too soon. Fermi, who equivocated on whether Majorana was dead or had deliberately disappeared, remarked that if he had killed himself, then with his family's religious sensibilities in mind, he would have taken care to ensure that no body was found.

Fermi's comment about Majorana's vanishing has been widely quoted as follows: "Majorana was too intelligent. If he has decided to disappear, no one will be able to find him. We have to consider all possibilities."[4] Not only did Fermi believe that Majorana would have hidden his plans had he committed suicide, but that Majorana could easily have arranged to disappear and covered his tracks in doing so. Whether he went to a monastery, as some supposed sightings led sleuths to conclude, or to South America, as other evidence might suggest, the mystery of Majorana's disappearance continues to fascinate historians even today.

A forensic investigation by Neapolitan physicist Salvatore Esposito has made a compelling case that Majorana fled by ship from Naples to Argentina.[5] There are several circumstantial pieces of evidence supporting this. Advocates of the deliberate disappearance theory point to the fact that Majorana had withdrawn all his savings from the bank just days before he was last seen. As to his destination, the timing and location of his supposed suicide coincided with the imminent departure of a ship from Naples harbour to Buenos Aires. This thesis has become amplified by several subsequent reports of a middle-aged man, a scientist or engineer with Majorana's features, sporadically visiting the university library there in the 1950s and 1960s.

These sober possibilities have been confused by more fanciful tales. These include that he was assassinated by Mussolini's agents or was kidnapped because of him being involved in the development of the atomic bomb. After examining the evidence, the author Leonardo Sciascia concluded that Majorana disappeared because he foresaw that nuclear forces would lead to nuclear weapons.[6] That Majorana was aware even in the mid-1930s of the potential implications of the work in nuclear physics is supported by him supposedly saying several times to the director of the Naples Physics Institute, Antonio Corelli: "Physics has taken a bad turn. We have all taken a bad turn".[7]

Edoardo Amaldi, who knew Majorana well through their time together as members of the Via Panisperna group, dismissed this on both a "human and historical level".[8] The "human" level dismissal is because Majorana was so fundamentally opposed to such an awful application of basic knowledge, he would never have let himself become part of such a venture. Supporting this is presumably Majorana's reference to the "bad turn" that physics had supposedly taken.

The "historical" critique is because of an anachronism, obvious once one immerses oneself in the chronology of this narrative instead of using knowledge from the (then) future: that Fermi's oversight on uranium fission would not be realised until 1939, a chain reaction would not be demonstrated until 1940, and not until 1945 would nuclear energy be released explosively. Majorana's disappearance in March 1938 predated all of these. There was nothing looming at the time of his vanishing to anticipate the apocalyptic vision of the atomic bomb test known as Trinity. Whereas the rush to exploit the energy of the atomic nucleus was very much in the air, at best the Joliot-Curies had at this juncture showed only how to modify quiescent matter to make it radioactive, while splitting atoms by machines such as at Cambridge took more energy to achieve than it liberated.

These and other breakthroughs that would pave the way to Trinity were still for the future. We will meet these in the next chapters, but even the beginning of this key sequence did not occur until several months after Majorana had disappeared. The idea that he fled because he saw the reality of H. G. Wells's fears as expressed in *The World Set Free* would only fit chronologically if Majorana had turned Wells's *The Time Machine* into reality and brought wisdom from the future back with him to 1938.

Or at least, this would be so if we only take account of what *experiment* had revealed by this stage. Serious speculation that it might be possible to "bring about transmutations of an explosive type" had, after all, been made by Frédéric Joliot-Curie in his Nobel Prize acceptance speech, in December 1935.[9] This is the point where Fermi's assessment of Majorana as a magician in *theoretical* physics might come into play.

Today, students of nuclear physics use the semi-empirical nuclear mass formula to answer questions like: what would happen when a neutron hits uranium? Doubtless Majorana, a supreme genius, was able to use that formula himself and might well have had the capacity to look forward from this to an atomic weapon. To do so, students use basic techniques in nuclear physics, which include concepts and formulae that would have been very familiar to him, not least as he himself had pioneered several of them. An integral part of this is to use Szilard's idea of chain reaction but applied to the heaviest of elements—uranium—rather than the light elements as Szilard had proposed.

Recall, there are two opposing forces at work within a nucleus: a short-range strong attractive force between neutrons and protons when they touch and a long-range electrical repulsive force among all the protons, wherever they are. Nuclear dynamics, the tension between stability and instability, is the result of competition between these two forces. A nuclear reaction occurs if you start with one force being dominant, add energy to enhance the competition, and then gain energy back as the other force takes over. This is analogous to pushing a waggon up and over a hill: a dynamo provides energy to drag it up the slope and then gravity takes over, speeding it down the far side.

It may be that Majorana's deep understanding of nuclear dynamics and the semi-empirical mass formula had shown him what happens when neutrons irradiate uranium and gave him a vision of the awesome implications: the added energy tips the balance when the electrical repulsion takes over, causing fragmentation into two smaller lumps, which then are pushed apart by these electrical forces and gain kinetic energy. And as fewer neutrons are required to make these smaller nuclei, there will also be neutrons released to repeat the sequence, leading to a chain reaction and an exponentially increasing release of explosive energy within a fraction of a second. It seems to me not impossible that Majorana, able to see further than even Fermi, let alone Szilard, anticipated theoretically by early 1938 what would only become apparent to mere mortals later that year. We may never know. What we do know is that something caused him to have a breakdown in 1934 and he became a recluse. Later recovering sufficiently to make a brief return to physics, he then disappeared again, this time for good, in 1938. With no corpse ever found, the popular theory that he fled to live in Argentina has survived, underpinned by Fermi's judgment that if Majorana had wanted to disappear he could.

If that was indeed Majorana's fate, the question is why did he flee? It is plausible that Majorana had foreseen the enormity of the atomic bomb and could not live with that apocalyptic vision. Nine months after Majorana's disappearance, Lise Meitner and her nephew, Otto Frisch, made the calculation that Majorana could easily have done before he vanished. Their conclusion, that a single neutron could liberate nuclear energy explosively from uranium, would change the world.

15

A Walk in the Woods

BOHR'S LIQUID DROP

Otto Frisch was a gifted pianist but rejected a career in music to become an experimental physicist. Based in Germany with a Jewish background, he watched the rise of Adolf Hitler and the Nazi party with horror. As the scourge of Nazism grew, Niels Bohr took on a rescue mission, trawling through German science centres where the Nazis were making savage recriminations against Jews and those who refused to swear allegiance to the party. Bohr was impressed by Frisch and offered him a position at the Institute of Theoretical Physics in Copenhagen, access to the laboratories, and an attic bedroom, its ceiling sloping down to the floor on one side.

Each week the Copenhagen group would meet to review their work and to discuss papers or news from elsewhere. At one of these an event occurred that would later prove seminal for Frisch.

Bohr had for some time been puzzled by Fermi's discovery that slow neutrons are so easily absorbed by the nuclei of some elements. "Slow" here is a relative concept. Fermi had indeed slowed the neutrons to a mere ten-thousandth of their initial speeds, but even at room temperature these speed-reduced thermal neutrons can travel about 2 kilometres in a second. This means that in a trillionth of a second the neutron could traverse a million nuclei. For one of those miniscule kernels to capture a neutron, the nucleus first must stop it, which involves absorbing its kinetic energy within a millionth of a trillionth of a second. Overall energy must be conserved so this kinetic energy must be transferred somewhere. There was no obvious way of getting rid of it in such a short time span, yet Fermi's measurements unambiguously showed that the neutrons were captured.

In 1936 at the weekly seminar the discussion centred on a paper by Hans Bethe, by then settled in the United States, who had some new ideas on uranium capturing neutrons after they had been slowed. It was while listening to this that Bohr had an epiphany. Members of the audience were momentarily worried because he suddenly sat still, his face completely dead. Some thought that he was ill. Then Bohr stood up from his seat and exclaimed, "Now I understand it!"

Up to then, the neutrons and protons in the nucleus had been regarded as interacting weakly and moving almost independently of each other. Bohr went to the board and explained his vision of the atomic nucleus. He saw the tightly packed cluster of protons and neutrons touching one another and interacting *strongly* as a coherent mass. In Bohr's model, an incoming neutron shares its energy with the entire ensemble, the nucleons so tightly gripped that none of the individual members has enough energy to escape. This excites the nucleus temporarily into an unstable compound state, which returns to a stable configuration once the reaction is over.

Like a cue ball in snooker hitting the reds, a neutron hitting a crowded nucleus gives up its energy to the nucleus's individual components. These recoil, bump into one another, and spread the impact around, sharing the energy among themselves. As a simple model, Bohr imagined these balls in a shallow basin.[1] If the basin were empty, a ball that was sent in would go down one slope and pass out on the opposite side with its original energy. However, when there are other balls in the bowl, the incident one will not be able to pass through freely. Instead, it will divide its energy first with one of the balls; these two will then share their energy with others, and so on until the original kinetic energy has been shared among all of them. In the real situation of a nucleus, this energy is manifested as heat. The nucleus becomes hot and then cools down by radiating gamma rays, but no individual constituent member escapes. He made an analogy with a liquid drop, held together by surface tension, which wobbles, deforms, and re-establishes its shape when gently stimulated, and applied this picture to an atomic nucleus responding to a slow neutron. Frisch remembered the occasion, and the model, clearly ever after. Within three years it would lead him to one of the most far-reaching breakthroughs in the history of nuclear energy.

A MISSED OPPORTUNITY

In 1936, a twenty-two-year-old Canadian named Leslie Cook graduated from the University of Toronto majoring in physics and chemistry. Cook had won a research fellowship to study for a PhD at the Kaiser Wilhelm Institute in Berlin, under the supervision of Otto Hahn. He would spend the next two years working alongside Hahn, Meitner, and their assistant, the chemist Fritz Strassmann, as they tried to disentangle the smorgasbord of radioactive isotopes resulting from the bombardment of uranium.

The same summer that Cook moved to Berlin, a Swiss chemist, Egon Bretscher, was at a conference in Zurich where he met Lise Meitner for the first time. She gave a talk on transuranic elements in which she summarised Fermi's work and explained how the Berlin team had come to similar conclusions. Meitner was claiming that she and Hahn had found elements heavier than uranium. She particularly talked about one that had properties like barium, and that is what Bretscher hooked onto as a chemist. He didn't believe that such an element could be. His chemical intuition immediately told him that this interpretation of being "like" barium was not correct, and he suspected that it must be nothing other than barium itself, but he didn't say anything in the discussion because Meitner "seemed to be rather touchy about things and he didn't want to openly attack her".[2]

Bretscher was based at Cambridge, part of Rutherford's team. Upon returning from the conference, he asked the nuclear physicist Norman Feather to make an experiment. Feather was involved in studying the way electrons jumped in the outer regions of different elements, and Bretscher said to him, "Here you have the ideal tool to find out whether that transuranic element is there, whether you can find the right jumps". Feather didn't take any action and Bretscher got very frustrated. For a long time, he said, "If only he had the sense to do something that really gives an answer to something!"[3] Had Feather followed Bretscher's urgings in 1936 they would have confirmed that the transuranic element was not there and that the barium must indeed be barium.

In Berlin, Cook's fellowship was coming to an end. He won a position at the Cavendish where, in the autumn of 1938, he would begin research with Bretscher.[4] During the summer Cook visited Cambridge to find lodgings and prepare for his impending transfer. He met Bretscher who quizzed him on what progress Hahn and Meitner had made in the last two years. Bretscher explained that he didn't believe in transuranic elements and that Hahn's "like barium" must be none other than barium itself. Cook then returned to Berlin for his remaining few weeks in the city and told Hahn about this conversation. When he told him Bretscher insisted that what Hahn believed to be transuranic was actually barium, Hahn retorted, "Well, you don't mean to say that I'm such a bad chemist that I don't recognise barium when I've got it in front of me?" That at least is what Bretscher heard back from Cook when he returned to work at Cambridge that autumn. Nothing more was heard after that from Hahn until a few months later.[5]

MIGRATION

While Frisch had left Germany for Denmark and escaped from Nazi persecution, at least for now, his aunt remained in Berlin. A growing number of Jewish scientists were leaving mainland Europe, Rudolf Peierls by 1935 being in the United Kingdom and many having crossed the Atlantic to the United States. Meitner had so far managed to escape the direct attention of the Nazis thanks to her Austrian background, but when Hitler invaded Austria in March 1938, she changed overnight, technically, from an Austrian into a German and became subject to the racial laws.

Through the summer, she continued working with Hahn and Strassmann while keeping to the background in hope of escaping notice. She was out of luck. Nazi administrators responsible for the Institute soon ordered her to wear the yellow Star of David, always. Hahn protested on her behalf without success. There were rumours that scientists might soon not be allowed to leave Germany, so she was urged by friends to escape at very short notice. In the autumn of 1938, she accepted an invitation to work in Stockholm at the Nobel Institute.

So ended a thirty-year-long collaboration with Otto Hahn in Berlin. Hahn continued their uranium studies with the help of Strassmann. The two chemists missed the physics expertise of Meitner but from Sweden she communicated by letter with ideas about interpreting the strange radioactivity produced by uranium bombardment. Just as Meitner left for Sweden, Hahn made a new tentative identification, this time that one of the substances behaved chemically like radium. Meitner the physicist was sceptical that radium, which is four places before uranium in the periodic table, could be made simply by the impact of a neutron. She wrote from Sweden to Hahn stressing that the result was incomprehensible, urging him not to publish it until he was completely sure. Hahn and Strassmann consequently decided to carry out more thorough tests to be quite certain what these substances were.

That same autumn the Nobel committee awarded the physics prize to Enrico Fermi for his discovery of transuranic elements. This was ironic given the amount of uncertainty still present among both the French and German teams who were continuing to investigate this phenomenon, along with the scepticism in Cambridge. More pertinent, perhaps, is that Niels Bohr discreetly alerted Fermi to the likelihood of the prize being awarded to him in order, first, to know whether as a citizen of fascist Italy Fermi would be allowed to accept it and, second, to allow Fermi, whose wife had a Jewish background, to make plans should he wish to escape from fascist persecution. Fermi and his family duly arranged to go to Sweden to receive the prize and then, instead of returning to Italy, to continue onwards to the United States. In total secrecy American scientists were alerted that Fermi was coming. After Enrico received the prize in December 1938, the Fermis crossed the Atlantic to settle initially in New York where the position of professor of physics at Columbia University awaited him.

Just a week after Fermi received the Nobel Prize, Hahn and Strassmann concluded their tests. They agreed with Meitner's scepticism. Following further chemical tests, they confirmed that the substances in their petri dishes were not radium. They had routinely added barium to help the chemical separations and been unable to distinguish the chemicals from barium. They concluded—as Bretscher had argued since 1936—that they were indeed isotopes of barium. How much Cook's report of Bretscher's opinion had made Hahn go back to look again and settle on barium we may never know.

Barium is an element in the middle of the periodic table with a nucleus about half the size of uranium. Moreover, at position 56, barium is but one place removed from lanthanum, number 57, which Irène Curie and Pavle Savić had seen and dismissed in their experiments in 1937.[6]

On 19 December Hahn wrote to Meitner with the news that he found barium, while also expressing his concern that this discovery seemed to violate all known physics. Recall, all evidence up to then was that any radioactive element transmutes into a more stable form that is either an immediate neighbour or at most next but one to it in the periodic table. All elements heavier than lead decay radioactively, in small sequential steps, eventually ending up as lead, the heaviest stable element. It was as if these heavy elements were positioned like floors in a high-rise, with lead on the ground floor, to which they could descend one or two floors at a time. What Hahn appeared to have found, however, was that the impact of a neutron could convert uranium into bismuth, at position 56, which in our analogy is deeply subterranean. Hahn, the chemist, decided to ask Meitner, the physicist, for advice on what this could mean. Meitner told him that she was about to leave for a vacation with her nephew, Otto Frisch. She explained that Frisch, who was based in Copenhagen, was travelling to Sweden to join her. Nonetheless, Meitner promised that she and Frisch would try to understand the implication of Hahn's news and gave Hahn her temporary address. Hahn began to write a paper for publication.[7]

"IT'S CALLED FISSION"

Otto Frisch traditionally celebrated Christmas with his aunt in Berlin. In December 1938, however, they were both émigrés. Frisch took a ferry from Copenhagen across the Øresund to Sweden, and then continued 150 miles north by train to Gothenburg and the nearby small resort town of Kungälv. The next morning when he came out of his hotel room, he found Lise Meitner at breakfast studying the letter she had received from Hahn. Frisch immediately noticed that she seemed worried by it.

When Frisch read Hahn's letter, with its assertion that uranium had somehow spawned barium, he found it "startling".[8] Frisch's surprise, and indeed initial scepticism, matched his aunt's. Individual protons or alpha particles were the largest fragments ever chipped from a nucleus; Hahn and Meitner's experiments had not provided enough energy to break off larger chunks.

After breakfast they left the hotel and set off on a walk through the snow-covered woods, Frisch on skis, Meitner in strong boots mounted on snowshoes. As they followed trails through the heavy snow, they argued about Hahn's results. The claim made no sense, and Frisch felt the simplest solution was that Hahn was mistaken.

They found a fallen tree and paused for a snack and a rest, sitting on its trunk. Meitner reminded her nephew that she had worked with Hahn for thirty years, that Hahn was probably the best radiochemist in the world, and that if he said so positively that he had found barium, then barium it must be. Meitner and Frisch had a conundrum.

They were agreed that it was not possible to have chipped numerous protons or alpha particle fragments from uranium as the energy to do so would have been far greater than the neutrons that Hahn had been using. The possibility that the nucleus had been sliced in half also seemed out of the question, Frisch remembering Bohr's moment of insight at the Copenhagen seminar three years earlier, when he had theorised that an atomic nucleus is like a very dense liquid drop, not a brittle solid. The forces of air resistance deform raindrops, which wobble and change shape, but surface tension resists the division of any drop into smaller ones. In Bohr's model, a similar behaviour would occur for nuclei: the impact of a relatively low energy neutron may cause the nucleus to vibrate or even to deform, but the strong forces among its constituents will resist its division into smaller parts.

The reason surface tension works against splitting a big drop into two or more droplets is because the surface areas of small objects are relatively larger than those of sizable ones. To see this, suppose you halve the radius of an object; its volume will go down eight times whereas the surface area will drop only by four. The ratio of surface to volume will therefore be twice as big in the smaller case than it was at the start. Minimising the surface area makes for stability, which is why two small drops of water coalesce into one larger one. In the case of the uranium nucleus, for example, splitting it into two smaller nuclei will increase the overall surface area. It takes energy to do this.

This much Frisch understood from Bohr's model. The first question was how much energy was involved. They found the answer to be of the order of thirty times more than one of Hahn's neutrons could provide. That is why at first sight the idea of the nucleus splitting seemed impossible.

It was then that Meitner or her nephew—it was never established which—had the key insight: unlike raindrops, the nuclei of uranium are saturated with electric charge. The disruptive electrical repulsion among uranium's ninety-two protons reduces uranium's binding energy by more than one third. This is the moment where they had the insight that Majorana may have foreseen. The empirical fact that no nuclei bigger than uranium manage to survive electrical disruption was already a hint that uranium might resemble a very unstable drop, whereby the impact of a single neutron might pull the drop into a dumbbell shape with a narrow waist in the middle. The electrical forces between the two fat ends might then break the drop into two pieces. Their relative sizes would be roughly equal but ultimately determined by chance. The two mid-sized drops would be two nuclei of elements in the middle regions of the periodic table—such as barium (with fifty-six protons) and krypton (with thirty-six protons to balance the books). In the recesses of Meitner's brain now surfaced a memory of Irène Joliot-Curie's tentative belief that she might have seen lanthanum. Could lanthanum, with fifty-seven protons, accompanied by bromine, with thirty-five, be another example?

The mental picture was of a uranium nucleus effectively being a combination of a barium and a krypton nucleus (or of lanthanum and bromine), gripped by the nuclear force. Excited at the possibility that the evidence had been hiding in plain sight, Frisch and Meitner needed to calculate if indeed the disruptive effect of electric charges could destabilise uranium enough to break the glue of surface tension and liberate these inner packages.

Their problem may be easier to visualise in reverse: suppose you *start* with one barium nucleus and one krypton nucleus and try to put them together to make uranium. Initially they repel one another electrically, and you must fight against this until they touch one another, at which point the strong attractive nuclear force takes over and pulls them together. It *costs* energy to get over the electrical barrier, but energy is *repaid* when the nuclear force merges them as uranium. The question is whether you get out more energy than you put in. In other words, does the competition between these two forces coalesce barium and krypton into uranium or expel them as two independent bundles?

This was a nuclear and electrical analogy of our example of driving a wagon uphill and then letting gravity give it speed on the downslope. Having received enough energy to surmount the nuclear barrier and form a dumbbell, the two nuclei rush apart with increasing speed thanks to the force of electrical repulsion. There is the potential to gain energy by splitting uranium into two smaller nuclei.[9]

With the concept now clear in their minds, they could work out the numbers. They gathered some scraps of paper from their backpack and began to compute. In Carl von Weizsäcker's mass formula for nuclei, two terms described the competition between volume and surface, which determined how surface tension resists a split. They already knew that this surface tension legislated for stability of uranium; the term describing the disruptive effects of electrostatic energy from the multiple protons was the one they now needed to compute.

An individual proton electrically repels all other protons wherever they are within the nucleus. In uranium, for example, there are ninety-two protons in all and each one feels the electrical presence of the other ninety-one. The total amount of electrical energy is proportional to the product 92×91. Now imagine that the nucleus split into two equal halves. Within each of these the electrostatic energy will be proportional to 46×45, or roughly one quarter of what was there originally. Two of these identical bundles will give a total proportional to 92 × 45, approximately one half of the original; as energy overall must be conserved, what has happened to the rest, the "missing" 92 × 46?

By using the formula, Meitner and Frisch computed the magnitude and found the electrostatic energy released to be more than twice the amount of energy needed to overcome the glue of surface tension. So roughly half of this energy has been used up in ungluing the two fragments—uranium can indeed be split—and the rest becomes the kinetic energy of these two positively charged lumps as they mutually repel one another. To their astonishment, nephew and aunt computed the amount to be some hundred times greater than had been found in any previously known radioactive decay; Einstein's formula $E = mc^2$ implied it was as if one fifth of the entire mass of a proton had been transformed into pure energy. Here, at last, was the way to release truly substantial amounts of the energy hidden within the atomic nucleus.

Frisch realised that if their calculations were correct, with this amount of kinetic energy the fragments would fly apart fast enough that they should leave visible trails in a cloud chamber. He decided he would do such an experiment as soon as he returned to Copenhagen, and hopefully verify what he and his aunt had deduced by theory. First, however, they spent the rest of their vacation checking their explanation, performing careful calculations for the splitting into barium and krypton, and then drafted a paper. As for Frisch, he wondered what Niels Bohr's reaction would be when he told him.

MISADVENTURES

Frisch returned to Copenhagen in considerable excitement and keen to share his and Meitner's speculation with Niels Bohr. It was 6 January 1939, and Bohr was about to leave for New York by ship, where he planned to spend a few days at Princeton University before going to Washington, DC, for an international conference on theoretical physics. Frisch recalled later how Bohr was in a rush and had only a few minutes available, but hardly had Frisch begun to tell his news when Bohr struck his forehead with his hand and exclaimed: "Oh what idiots we have all been not to have seen this before! This is wonderful! It has to be like this".[10]

Bohr's reaction shows clearly how, with hindsight, this should have been seen long ago. (Given that after receiving the prompt from Hahn's experiment, Frisch and Meitner made a calculation that involved concepts well known to Majorana, it is not easy to dismiss the possibility that he had already seen this much for himself.) Bohr then asked Frisch if he and Meitner had written a paper about it, and Frisch replied: "Not yet, but we will at once". Bohr urged them to do so without delay and promised not to talk about their breakthrough before the paper was published. He then hurried off to catch his ship, accompanied by his scientific secretary and assistant, Léon Rosenfeld.

Frisch was aware that in biology there is a process in which single cells divide in two. The analogy with the splitting of uranium seemed apt to him and so he asked a biologist, William Arnold, for the name of this phenomenon. "Fission" came the reply. As a result Frisch coined the phrase *nuclear fission* and included it in the paper.

Lise Meitner had now returned to Stockholm, so the paper was completed after several long-distance telephone calls. These discussions with Meitner, framing and re-drafting their paper, took five days. Frisch had no sense of urgency because apart from Meitner and himself only Bohr knew of their theory and had promised to say nothing until the report was in print. Moreover, he felt no claim to originality, remarking that he "was just lucky to be with Lise Meitner when she received advance notice of Hahn's and Strassmann's discovery".[11] He wanted to demonstrate by experiment that fission can occur, and to do so by looking for fragments of uranium in a cloud chamber. The higher the energy of the fragments, the greater is their range through matter before being brought to rest. His and Meitner's computations predicted what energies these fragments should have, so by detecting them and measuring their ranges, he would be able to test the idea both conceptually—that fission is real—and quantitatively.

It took longer than expected. To photograph the tracks of fragments from uranium fission he had to ask the institute's glassblower to build a special piece of apparatus. Once he had the pieces to hand, he took a whole day to put them together and prepare for the experiments. It was not until the evening of 12 January that he finally began the tests.

He worked carefully, aware that physicists around the world would follow up and check his results. He fired neutrons at uranium and recorded bursts of fission fragments manifested as a flow of charge across his ionisation chamber. By 5:00 in the morning on Friday 13 January he was satisfied that he had demonstrated experimentally the reality of the phenomenon, which until then had been but theory. He spent the weekend writing a report about the experiments and completed it on Monday 16 January. Finally, he sent this account, together with the paper written with Meitner, to the British journal *Nature* in London.

Meanwhile, on the transatlantic crossing, Bohr was working out the detailed theory of fission and beginning to consider its implications. Bohr's style was always to think aloud and discuss his ideas in conversation with another scientist. Rosenfeld was a perfect foil as they walked the decks of the *SS Drottningholm* during the ten days crossing en route to New York. The ship was infamous for rolling, and had acquired the nickname "*Rollinghome*", but Bohr seems to have been unaffected as day by day he reviewed the evidence, deepening his understanding of fission and its implications. By the time of their arrival, Rosenfeld, who had been immersed in dialogue with Bohr, was also an expert; unfortunately, in his excitement Bohr had failed to alert Rosenfeld that he had undertaken to keep Frisch and Meitner's insight confidential until it appeared in print. Back in Copenhagen, six time zones later in the day, it was only now that Frisch was mailing the paper to *Nature*.

Upon arrival at the 57th Street pier in Manhattan, at lunchtime on 16 January, Bohr and Rosenfeld were greeted at the dockside by a collection of scientists, old friends wanting to greet the great physicist to North America. Among them was Enrico Fermi, who had been in the United States for only two weeks after receiving the Nobel Prize. After completing immigration and clearing customs formalities, Bohr and Rosenfeld were due to visit Princeton. Bohr however had business in New York which would take a day or two, and so Rosenfeld went on to Princeton alone, Bohr remaining in Manhattan. This was when the misadventures began.

It was standard practice at Princeton like many universities for visitors to give an informal presentation about news in physics, especially when they had come all the way from Europe. Unaware of Bohr's agreement to maintain secrecy, and mistakenly believing that Frisch and Meitner's work was published or at least imminent, Rosenfeld now freely reported the news of nuclear fission. He included Bohr's own detailed analysis that they had shared during the trans-Atlantic crossing. Inadvertently the greatest discovery in nuclear physics was gifted to the New World before anyone in Europe had seen a written report.

Meanwhile, the Princeton scientists were unaware that there was any embargo and immediately organised a meeting of leading physicists. By the time that Bohr arrived, a few days later, Frisch and Meitner's breakthrough had become public property. Bohr immediately told the Princeton scientists the history of the new phenomenon and asked them to respect the priority of Meitner and Frisch whose work had not yet appeared in print. However, the genie had escaped.

A major conference on theoretical physics was due to take place in Washington on 26 January, and on the eve of the conference one of the scientists who had heard the news from Princeton burst into Enrico Fermi's office sharing what Rosenfeld had leaked and Bohr had now confirmed. This was the moment when Fermi first learned of nuclear fission: that the bombardment of uranium by neutrons splits it in two, making lighter elements, not—as he had previously anticipated—transuranic ones. This appears to be one of the few occasions when Fermi was psychologically defeated. He mulled over what had happened in the last few years, recalling the range of unusual phenomena that he had seen in his data since 1934 and which now he realised he had completely misinterpreted.

16
Chain Reaction

"WHAT FOOLS WE HAVE BEEN"

The rise of fascism in Italy led to the breakup of Fermi's group. In 1937 its youngest member, the communist Bruno Pontecorvo, moved to Paris to work with the Joliot-Curies who had similar political leanings. Joliot was also friendly with Igor Kurchatov, a physicist in Moscow, who had been inspired by Fermi's work. They communicated by letter and read one another's published papers.

Kurchatov was an imposing figure with dark hair and an intense gaze. While the sides of his face were clean shaven, the front of his chin sported a remarkable beard which hung down onto his upper chest, like a table napkin. Kurchatov had discovered isomers, examples of nuclei in which a single isotope appeared to have different amounts of radioactivity. In Paris, Pontecorvo and the Joliot-Curies began to investigate the phenomenon.

They discovered that neutrons and protons in the nucleus can take on different energy states, as electrons can in atoms. Each rung on the energy ladder has a slightly different energy, and by the mass energy equivalence this implies a different mass. The difference in these masses is a mere one part in several thousand, which is too small to be measured directly, hence the sobriquet *isomer*, for what appeared to be "equal masses". Although trifling, these differences enabled photons to be emitted as the isomers tumbled down the energy ladder. It was by measuring the spectrum of these photons that the pattern of isomers was disentangled. The results further demonstrated that a nucleus is a rich collection of constituents, which can move, orbit, and vibrate relative to one another, these motions all controlled by the laws of quantum physics. Their mutual interest in isomers led to close contact between the team in Paris and Kurchatov in Moscow.

Meanwhile Frédéric Joliot-Curie, now thirty-seven years old, was appointed professor at the Collège de France, the most prestigious post in the nation. His eminence brought influence with which he persuaded the National Funds for Scientific Research to buy an old electrical plant in Ivry-sur-Seine, a suburb southeast of Paris, and convert it into a nuclear physics laboratory. His goal was to create radioactive isotopes at the laboratory for use in research. Irène remained with her own research team at the Radium Institute.

Hahn and Strassmann's paper about their experiments irradiating uranium with neutrons arrived in Paris on 16 January 1939. Frédéric immediately understood what must have happened. Irène had observed similar phenomena in 1937 when she and Pavle Savić had tentatively identified lanthanum, element 57, but had not been confident enough to confront criticism of the results and had recoiled from publishing. Frédéric now suspected she had been right all along and that her neutrons must have split the uranium in two. For several days the news was the hot topic of discussion throughout the French group. One member, Polish émigré Lew Kowarski, recalled how nobody talked of anything else. Irène raged at having missed out on getting credit for this discovery, for which she could have won a second Nobel Prize and emulated her mother. She told Frédéric "What fools we have been"—using, as Kowarski commented, "a somewhat stronger word".[1]

When Joliot wrote to Kurchatov about the phenomenon, the Russian immediately began to investigate whether fission also liberated any secondary neutrons. The Paris group did likewise, as did others in the United States and England. The critical question was whether more than one neutron was released for every neutron that caused the fission in the first place. If this was so, a self-sustaining chain of reactions could occur. This suggested that energy release could build up exponentially as a sequence of fissions spread throughout a lump of uranium.

A key feature of uranium is that its nucleus contains more than 140 neutrons, significantly more than is enough to satisfy the needs of smaller nuclei such as barium, krypton, or lanthanum, the likely debris. It was plausible then that, during the fission of uranium, some neutrons would also be liberated. The moment he heard the news, Frédéric immediately devised a simple experiment to detect the neutrons—and failed. The problem was that he was irradiating the uranium with a source of neutrons so intense that his attempt to identify additional particles was like trying to detect a rain shower while standing beneath a waterfall.

Undeterred, he spent the next few days designing a new experiment, one that would look for evidence of radioactivity in the debris when the neutrons hit the uranium. The method of doing this was remarkably simple and testimony to his skill as an experimental physicist.

First, he engineered two brass tubes, one of which he coated with uranium. Next, he made some Bakelite cylinders, larger than the tubes so that he could surround them like napkin rings. He placed a neutron source inside the brass tube that was free of uranium and put the source and the tube inside the Bakelite ring. After a few minutes of irradiation, he removed the ring, took it to a Geiger counter, and confirmed there was no radioactivity. That done, he repeated the exercise, but this time using the uranium-coated brass tube. As before, he surrounded the tube with a Bakelite ring. When he removed the ring on this occasion, and took it to the Geiger counter, it set the device clicking. This proved that radioactive fragments of uranium had adhered to the Bakelite, proving that the uranium had been shattered.

Frédéric repeated the test with rings of various sizes and confirmed that the intensity of the radiation was less for large rings than for small ones, indeed that with the largest ring there was no radioactivity at all. The bigger the ring, the further away it was from the uranium. As the intensity fell in proportion to this distance, he concluded that the radiation was undoubtedly the result of the uranium and not some other source.

So far, he had demonstrated that the uranium had split and produced radioactive products. His technique was different from Frisch's, but his results were the same. The challenge now was to be able to detect the liberated neutrons. He and Kowarski found a clever way to do this.

By performing calculations of the energy released in fission, similar to those Frisch and Meitner had done, it was possible to estimate the energy of any neutrons that were released. It turned out these would be much faster (higher energy) than the neutrons that Frédéric was using to irradiate the uranium. The trick that he devised was to irradiate the uranium with slow neutrons and to use a detector that was sensitive only to fast ones.

He and Kowarski began this experiment during the last week of February, and collaborated with Hans Halban, a self-aggrandizing, somewhat arrogant physicist of Austrian-Jewish descent, born in Leipzig. It took them about six weeks until, in April, they finally were convinced that they had demonstrated that the fission of a uranium nucleus liberates more than one neutron, most probably two. The implications of this discovery were nothing short of awesome. The possibility that these freed neutrons could initiate further fissions, producing a self-sustaining nuclear reaction, was now realised.

Consider for a moment the implication of one neutron splitting a uranium nucleus into two chunks, liberating energy and two neutrons. There is a chance that these two neutrons will hit two additional uranium atoms and repeat the fission. If this happens in subsequent collisions, there will now be four neutrons freed to make four fissions, leading to eight, sixteen, and so on—the number of neutrons doubling at each step. It would only be necessary to irradiate uranium with a few neutrons to set up reactions that would continue spontaneously until all the uranium was used up. This created the potential for an immense release of energy.

On 22 April, Joliot was satisfied that his team had established that a chain reaction was possible, and in the first week of May he applied for three patents. Two dealt with the potential application of fission to nuclear power, and the third, which was secret, related to explosives. On 8 May, he went to Brussels to negotiate the acquisition of uranium stock from the Belgian Congo with a view to building an explosive uranium device in the French Sahara.[2]

THE REACTION CHAIN

The prospect of getting usable nuclear energy from uranium excited scientists around the world. That a nuclear chain reaction might also lead to a catastrophic new form of explosive, qualitatively different to anything that had gone before, was obvious to any scientist who had read the papers now readily available in the scientific literature.

In the United Kingdom scientists alerted the government to uranium's strategic importance. Ernest Rutherford had died suddenly in 1937 at the age of sixty-six and did not live to see the climax of the nuclear energy adventure that he and his colleagues had instigated. Fermi, in the United States, immediately realised fission's implications. In Germany there was a similarly intense response: all reference to atomic energy and uranium reactions was immediately censored in the German media.

Bohr had brought the news of Frisch and Meitner's breakthrough to the United States, if inadvertently, but was yet unaware of Frisch's follow-up experiment confirming fission by detecting its nuclear fragments and measuring their energies. This next step was so obvious that almost as soon as they heard Bohr's news, several physicists in North America conducted similar experiments themselves.

Fermi irradiated uranium with neutrons and detected its fission fragments by using an ionisation chamber, his experiment identical in concept to Frisch's but for the method of detection. Leo Szilard, who had also fled Europe to North America, immediately resurrected his fission obsession, translating it from his original mistaken focus on light nuclei to the heaviest reaches of the periodic table, uranium. He visited Washington to see Edward Teller, like him a Jewish-Hungarian émigré physicist. Teller's memoir recalled their conversation:

Szilard said: "You heard Bohr on fission?"

"Yes", I replied.[3]

Szilard explained the urgency in determining whether a splitting nucleus would also release more neutrons and continued: "You know what that means! Hitler's success could depend on it."

The main purpose of Szilard's visit was to lobby Teller to join him in a campaign to keep these developments secret. Such a hope was naive. Hahn and Strassmann's experiment had already been published in January and now Frisch and Meitner's explanation of their results was in the public domain. Fermi and others had immediately seen the potential implications of a chain reaction and were even now setting up experiments to see if fission produces secondary neutrons, like droplets accompanying the bigger drops of barium and krypton. There was no reason to think others around the world would not do so—as indeed back in Europe Joliot-Curie already was.

Fermi's idea was simple. In place of an ionisation chamber as detector, he used foil made of rhodium, a precious silvery white metal that becomes radioactive when hit by neutrons. The half-life of this radioactivity is just 45 seconds, which showed that neutrons had hit, and the rhodium settled back to its ambient state within minutes.

For his neutron source, he mixed radium with beryllium, in the centre of a small hollow container. Aided by a young collaborator, Herb Anderson, Fermi sealed it and placed it into a metre-deep tank of water. The neutrons poured out from the source and slowed as they passed through the water. He placed his rhodium detectors at varying distances away from the source. These verified that neutrons were produced, as of course he knew, but the importance here was the rate, so he first measured the ambient neutron signal in the absence of uranium. Next he added uranium oxide, a yellow powder. The neutrons now blasted through the uranium, creating fission and, potentially, further neutrons.

To his satisfaction, the detector showed a higher rate of the neutrons than before. This implied that when neutrons hit uranium, the fission also liberated secondary neutrons. Fermi did some calculations and concluded that about two neutrons must be produced in uranium fission for every neutron absorbed by the uranium and permanently lost. Unknown to him this agreed with what the French team was finding. A chain reaction looked likely, although there remained considerable uncertainties. If one fission instigated by one neutron had liberated just one neutron, the reaction would eventually die out because some neutrons will escape without hitting anything or be absorbed by uranium atoms without inducing fission. The latter, after all, is what Fermi had erroneously believed to be the whole story when he had been searching for transuranic elements. Now his focus was on fission and the number of secondary neutrons. Szilard performed a similar experiment, and the trio joined forces in their publication.[4]

The possibility remained that alongside the drama of fission, some neutrons could be captured by uranium and produce transuranic elements. At the time, however, the chain reaction was the centre of Fermi and everyone's interest. The possibility that transuranic elements might also be produced by neutrons that did not create fission had slipped momentarily from their attention.

WHY ISN'T EARTH'S CRUST EXPLODING?

Hardly had the proof arrived that a chain reaction can occur before a question immediately struck scientists. It was this: if a chain reaction can indeed liberate energy explosively, why are the rocks around us, which contain uranium and are being hit continuously by cosmic rays, neutrons, and other sources of radioactivity, not liable to detonate spontaneously?

The answer came from two steps. The first was an empirical puzzle involving an intriguing anomaly in some of the results on the fission of thorium, the neighbour next but one to uranium in the periodic table. Whereas at first sight the fission products of the two were very similar, the way that neutrons produced the reaction turned out to be different. The second was how this anomaly gave Niels Bohr the clue that solved the puzzle.

To understand the empirical anomaly, here is an analogy due to Rudolf Peierls. Peierls was the young theorist who with Hans Bethe in 1934 had developed Fermi's model of beta decay into a viable theory of the neutrino. He had many talents; one was as a pedagogue.

Peierls's analogy was between a neutron hitting uranium and a ball thrown from outside a house hitting a glass window. The ball might shatter the window, or bounce off, or if the window happened to be open, would pass through into the room behind. The window is the analogue of uranium, and the ball plays the role of a neutron. What happens next cannot be predicted with certainty but given enough throws can be predicted in the form of probabilities. There are three possibilities; experiment can determine their relative likelihoods.

The first option—shattering the window—corresponds to a neutron shattering the nucleus: fission. The second, bouncing off the glass leaving it unbroken, is known in the physics jargon as elastic scattering.

This is like the neutron bouncing off the uranium, leaving both uranium and neutron unharmed; the neutron is now free to try its chances again with another nucleus in its path or to escape from the ensemble if it meets none. The third, if the window is open and the ball enters the room, is like the uranium nucleus capturing the neutron and becoming momentarily a collection of 239 neutrons and protons. If this happened followed by beta radioactivity, which converts a neutron into a proton and increases the atomic number by one, it would convert the original uranium nucleus into that of a transuranic element: neptunium with atomic number 93. Were the neptunium itself to beta decay, it would convert into plutonium, with ninety-four protons.

Physicists measured what happened when neutrons hit uranium and worked out the relative chance of each of these occurring. They found it depends upon the neutron's energy—by analogy, how fast the ball was thrown at the window. It turned out that low energy neutrons with a kinetic energy of a mere 25 electron volts had a bigger chance of being captured. This energy corresponds to about one thousand times the agitation of particles at room temperature. Below this energy, thermal neutrons bounce off the uranium.

Slowing the neutron to such energies, by scattering from water, paraffin, or graphite for example—moderating them, in the jargon—was therefore probably a route to production of transuranic elements. As Fermi's experiments in Rome had first revealed the efficacy of moderation and then systematically applied it to elements as heavy as uranium, he had probably indeed made some neptunium and plutonium, though these were outnumbered by the noise of the yet unrecognised fission, which can happen at all energies.

Now to the enigma in the data that inspired Bohr to his explanation.

When Hahn, Meitner, and Strassmann had performed the experiments that eventually led to the discovery of nuclear fission, they had bombarded both uranium and thorium. They found similar anomalies in both cases. After Frisch and Meitner had their insight that the uranium results were due to fission, in his experiments in Copenhagen

Frisch confirmed neutrons split thorium as well as uranium. However, he found a difference between the two. When he first passed neutrons through paraffin, slowing them down, the fission rate increased for uranium, whereas for thorium it died away. The fission probability in uranium occurred for both slow and fast neutrons. For thorium, however, it only occurred for fast neutrons.

In the United States, at the end of January and yet unaware of Frisch's experiment, a group at the Department of Terrestrial Magnetism in Washington, DC, used a Van de Graaff accelerator to spawn neutrons of different energies. They independently made the same discovery as Frisch, but with added information: they could vary the energy of the neutrons and determine quantitatively how the phenomena behaved as a function of that energy for both thorium and uranium all the way from the lowest energy neutrons to the highest. They too found that while fast neutrons can split both uranium and thorium, slow neutrons only do so with uranium and, moreover, with much greater probability.

They concluded that fissions of uranium by fast and by slow neutrons are produced by different processes. This was a clear empirical fact, but why should nature behave this way? This was the question that Niels Bohr now confronted.

In hindsight, his brilliant insight was remarkably simple. He made use of a discovery about the nucleus that, if it had been noticed by others at all, had been regarded as little more than taxonomy. Bohr now thrust it into centre stage.

A Canadian, Arthur Dempster, had been filtering isotopes in a mass spectrometer in which a magnetic field deflects ions, lighter ones more easily than heavier ones. Ions of a single element contain various isotopes, which are differentiated by their masses. The magnetic field separates these isotopes such that their relative positions when they hit a detector give a measure of their relative masses and the intensity of the signals reveals their relative abundance. Dempster found rare isotopes of several elements this way. What would turn out to be his greatest discovery was of a rare isotope of uranium, uranium-235, with three fewer neutrons than the common form, uranium-238. What initially appeared to be just another isotope for the collection would prove key to the explosive release of nuclear energy.

Thanks to Dempster, Bohr knew that natural thorium is made of but one isotope—thorium-232, or Th232—whereas uranium occurs in two forms—U238 and U235. Unlike Th232 and U238, U235 contains an odd number of nucleons. Odds and evens, and the chance of a neutron hitting U238 or U235, inspired Bohr's insight. The extra stability of even numbers relative to odd, which Ettore Majorana had used in his first theory of nuclear structure, causes fission to be more likely in an isotope with an odd number of constituents, such as U235, than in one with an even number, such as U238. In natural uranium, U235 is only seven in one thousand atoms. The dominant U238 acts as a blanket that covers any nearby U235, making a succession of fissions rare and the chance of a chain reaction negligible. Only if the neutrons encounter some of the rare isotope U235 before exiting the uranium target will fission occur. Thus, the fission of natural rocks is rare because they contain so little U235.

While that is good news for our daily affairs, it showed that it would be difficult to extract nuclear energy from raw uranium, let alone make an explosive device. Bohr's analysis exposed the naivete of many physicists who, upon learning of fission, prematurely foresaw a nuclear explosion. One of the best known of these is J. Robert Oppenheimer, a lean chain-smoking American polymath and theoretical physicist based at the University of California, Berkeley.

Within a week of hearing the news, there was a drawing on his office blackboard of a bomb. This has led to some modern-day descriptions of Oppenheimer as the "father of the atomic bomb" which, as we shall see, is wrong on many fronts. Not least, this early drawing was described as "very bad, an execrable drawing of a bomb", and but for this sketch, Oppenheimer's reaction seems to have been no different to that of many physicists, such as Szilard, Teller, and even Fermi.[5] Only

Bohr realised that the stability of uranium rocks implied that chain reactions leading to explosions are not the natural order, and then analysed the problem to understand why.

Yet atomic explosions are possible, as is today self-evident. Oppenheimer would become, if not the father, the "midwife" of the atomic bomb. For there exists a consequence of Bohr's analysis, though surprisingly no one at first realised it.

Neutrons are the spark that can light the nuclear fire, but the flame is fickle. To liberate nuclear energy effectively requires *enrichment*—the increase of the relative amount of uranium-235 in the target—and lots of uranium. Bohr published his insight in the American journal *Physical Review* on 1 September 1939, the same day that Germany invaded Poland and the Second World War began.

At this critical moment in history, an atomic bomb small enough to be delivered by a plane or gun seemed out of the question. Instead, the scientific community expected that if several tonnes of uranium were piled together and irradiated with neutrons, it might be possible to liberate energy and produce power for industry. This is in essence what today is known as a nuclear reactor; in 1939 it was but a theoretical conjecture that remained to be proven.

THE ODDS OF FISSION

Nature does not always give up its secrets easily. Bohr alone seems to have puzzled about the enigma of uranium's stability under bombardment by cosmic neutrons. Any number of people, had they paused and thought, could have done what he did. As Majorana and Heisenberg had known all along, and by 1939 was known to the small but growing number of nuclear physicists, neutrons in the nucleus like to clump in pairs, as do protons. The result is that nuclei containing an even number of nucleons, like thorium-232 or uranium-238, are more deeply bound and relatively more stable, nucleon for nucleon, than those with odd numbers, like uranium-235.

Bohr had understood in a flash the implication for fission. Tightly bound, even-numbered isotopes like U238 need a powerful hit to distort them into a dumbbell shape before the electrical repulsion between the two bulges initiates fission. The single invader neutron that has converted the tightly bound, even-numbered isotope momentarily to a less stable, higher energy, odd-numbered one first requires a gift of that energy. This must be supplied by the incident neutron itself. The result: fission of U238 is caused by *fast* neutrons.

In marked contrast, an odd-numbered starting nucleus has no need for an energy down payment. In U235, a neutron of *any* energy will pair with the isotope's lone singleton and initiate fission; therefore, fission of U235 occurs even with very slow thermal neutrons. In a nuclear nutshell, U238 needs energy input to be elongated enough to fission, whereas U235 gains this energy merely by picking up a neutron, *any* neutron, and pairing it off. Fission of U235 happens at all energies, for neutrons fast, slow, or thermal.

Bohr made one further observation about neutrons being captured by uranium. U235 does not capture neutrons; U238, however, captures them copiously if their kinetic energies are around 25 electron volts. Neutrons of this energy are thus trapped and the resulting unstable nucleus beta decays to make first neptunium-239 and then plutonium-239.

Everyone initially had focused on Bohr's explanation of why fission of natural uranium is not explosive. The relevance of the capture phenomenon for U238 opened another route to releasing nuclear energy when it was realised that plutonium-239, once made, should liberate fission energy even more easily than uranium. At least, in theory. Only experiment could show if this were really the case.

PART IV

NUCLEAR SECRETS
1940–1960

17
"Extremely Powerful Bombs"

During 1939 the news of nuclear fission and the possibility of a chain reaction created for many scientists a reflex action of escalating paranoia. The apparently logical sequence was this: fission of uranium liberates a large amount of energy; a chain reaction will liberate that energy exponentially; this will lead to a nuclear explosion.

It was increasingly obvious during the year that the Nazis had military designs on Europe. The scientists' paranoia was not helped in June when Otto Hahn's assistant, Siegfried Flügge, wrote in the German journal *Die Naturwissenschaften* about the possibility of liberating nuclear energy and creating a "violent explosion". In the United Kingdom, C. P. Snow, the editor of *Discovery* magazine, on 1 September reflected on the possibility of "nations dropping bouquets of uranium bombs" with the ominous conclusion, "if it is not made in America this year it may be next year in Germany". He added that if there is nothing in the physical laws to prevent a nuclear explosion; it will "certainly be carried out somewhere in the world".[1] Snow's article was not widely noticed, however, as attention that fateful day was focused on the start of the war.

In America during August, the highly vocal Leo Szilard had acted following principles like those of Lewis Carroll's Queen of Hearts: conclusions first, analysis later. Notwithstanding that his paper with Fermi and Anderson admitted "more information . . . [is] required before we can conclude whether a chain reaction is possible in mixtures of uranium and water", he managed to convince Albert Einstein to write to President Roosevelt, alerting him that fission plus a chain reaction opens the route to the atomic bomb.[2] Large amounts of power could be produced by a chain reaction, and by harnessing this power, "extremely powerful bombs" were conceivable.[3]

Fermi and Bohr were more sanguine. Fermi assessed there to be "little likelihood of an atomic bomb". He added that he judged the bombers were "pursuing a chimera".[4] As for Bohr, his detailed analysis had shown that in the absence of being able to produce a significant enrichment of uranium-235, an atomic explosion was not viable, and he published that result the same day as the war began.

Folk wisdom is that Einstein's letter kick-started the Manhattan Project and led inexorably to the atomic bomb. That is far from what happened.

Szilard seemed unaware that Bohr had demonstrated theoretically that natural uranium has no explosive power, as he never alerted Einstein to this key information. As for Einstein, his glory days of relativity were past and he was becoming a peripheral figure, reluctant to accept the new quantum mechanics. When Szilard outlined the idea of a chain reaction, this was news to Einstein who, unaware of Bohr's well-argued scepticism, duly signed Szilard's letter.[5]

Reactions to these urgings in America were somewhat lethargic. Not until October was there a response, which was the recommendation to set up a Uranium Committee consisting of a handful of scientists. One of its first decisions was to authorise purchase of 4 tonnes of graphite and 50 tonnes of uranium oxide to support experiments that Fermi was doing at Columbia University.

Fermi wanted to see if a self-sustaining fission chain reaction was possible in a large amount of natural uranium. From his experience at Rome, he knew that slowing the neutrons increases the likelihood of reaction. Instead of using water he planned to use graphite to moderate them. His plan was that after slowing the neutrons this moderation would increase the chance of making further fissions, which would in turn maintain the chain reaction. A pile of graphite mixed with a vast amount of natural uranium might then produce enough secondary slow neutrons to keep the reaction going.

This was all theoretical. In any event, there was no explosive bomb here; 50 tonnes of uranium was far too much to be an explosive mixture deliverable in warfare. As Bohr's paper implied, there would be need for separation of uranium-235 or enrichment of natural uranium to make an energy-producing machine on a small enough scale, let alone an explosive device. The Uranium Committee agreed that uranium-235 enrichment would be needed if the fast liberation of energy were ever to be achieved. The question then was how best to do this.

One approach was to use a high-speed centrifuge. A cylinder spins rapidly on its vertical axis and separates a gaseous mixture of U235 and U238, thanks to their small difference in masses. The lighter U235 is less affected and can be drawn off at the top of the cylinder. A cascade of hundreds of centrifuges then would hopefully produce a rich mixture. Another approach was to use gaseous diffusion, where molecules of the lighter isotope pass more easily through a porous membrane. Here again, repetition of the diffusion through membranes hundreds of times would eventually increase the percentage of U235 over U238. In any event, one thing was certain: the days of bench-top science or even of machines in purpose-built laboratories were gone. Enriching uranium to liberate energy at all would be an industrial enterprise; a military explosion was not yet on anyone's serious agenda. By June 1940, all work on uranium research in the United States was classified top secret.

THE FRENCH CONNECTION

One version of history is that nuclear fission was not discovered by Hahn and Strassmann. The opinion in the Joliot-Curies' lab was that Germanic chemistry had done no more than extend what Irène and her assistant Pavle Savić had reported the previous year, and that neither Hahn nor Strassmann understood its significance any better than had Irène at the time. The true discovery of fission was made by Frisch and Meitner on a tree stump in a Swedish forest.

Following fission's announcement, Frédéric had sprung into action with his demonstration that it would be possible to make a chain reaction. With his colleagues, he rushed out a paper to show they were first; this was to be a prize for France, the Curie name was still on display, and French atomic science was back in the limelight.

Frédéric now planned to build an experimental test reactor to see if a chain reaction would take place on a large scale. As 5 tonnes of uranium would be required, the operation was moved from the Collège de France to the more spacious laboratory at Ivry-sur-Seine in the Paris suburbs. This project required financial support, so during the summer of 1939 as war loomed, he performed demonstrations to impress the French ministry of supply rather than pursuing scientifically impeccable proofs.

His publicity campaign worked as *Time* magazine included Frédéric on its front cover that autumn. His team's actual scientific paper was also their last open publication before secrecy enveloped nuclear physics worldwide. Frédéric's publicity was very successful, but the paper was less impressive in the opinion of many scientists; when Fermi was asked what he thought of it, he supposedly replied: "not much". One of Frédéric's collaborators admitted later: "Scientifically the paper was not very impressive. As a demonstration it was".[6]

Frédéric's team planned a pragmatic series of experiments to convert atomic energy from an arcane idea to a practical power source that could be switched on and off at any time. Their first attempt to build a nuclear reactor had the uranium in small blocks all inserted inside a graphite shield, which was planned to be a moderator for the neutrons. It didn't work, as both the initial and the secondary—fission-produced—neutrons needed to be slowed, which could only be done within the uranium mix. They doused the uranium in water, but that didn't help either as it absorbed too many of the precious neutrons. The reason is that the hydrogen in normal water, H_2O, captures neutrons; in so-called heavy water, D_2O, the hydrogen is replaced by deuterium, whose nuclei already contain a neutron. Deuterons have little affinity to capture neutrons, and so heavy water is a much more effective moderator. So, Frédéric decided to use heavy water. He also redesigned the experiment.

He proposed to build a hollow sphere filled with heavy water into which uranium oxide powder was dispersed. The sphere was mounted on a spindle, enabling it to rotate and prevent the powder from settling. At this stage his goal was to show that it was possible to extract energy from uranium inside the device using heavy water as the moderator. As for the heavy water, the only industrial source was the Norwegian hydroelectric power plant Norsk Hydro at Rjukan. The go-between was Jacques Allier who worked for the Banque de Paris, which had invested in the Hydro plant; Allier's covert work, however, was in the French Secret Service.

In March 1940, Allier flew to Oslo where he was joined by two agents from the French embassy. There was already suspicion that Germany was on the same path and they were under strict orders that the heavy water must not fall into enemy hands. These suspicions were confirmed when, upon telling the factory that he wanted to purchase all the heavy water that they had, he was informed that there were 165 litres, stored in aluminium cans and that the Germans were already bidding for it.

Allier arranged for these cans to be packed into two wooden cases and asked for two further identical cases to be prepared. This additional pair would be a decoy, filled with ordinary water. Allier was aware he was being watched, so he went to the airline office and booked a seat for Paris along with two cases of luggage—the fake canisters containing ordinary water. Meanwhile a chartered DC-3 private jet at another airfield was secretly loaded with the real heavy water.

Allier boarded his plane, but when it paused briefly at the end of the runway before takeoff, the pilot opened the door so Allier could get out. He was driven to the other airfield and accompanied the heavy water to Scotland from where it was successfully transported to Paris. The precautions proved critical because the commercial flight from Oslo to Paris was intercepted by German bombers and forced to land near Hamburg, where the decoy water canisters were unloaded. On 9 April 1940, Nazi Germany invaded Norway, but by then the world supply of genuine heavy water was safely in Frédéric's laboratory.

It would not long be out of reach, however, for in May Germany invaded France and by early June Nazi troops were at the gates of Paris. Frédéric first moved the heavy water 260 miles south by truck to Clermont-Ferrand in hope of continuing reactor development there, but on 17 June Allier brought orders that it be transported to England for better security. Halban and Kowarski accompanied it from Bordeaux to Falmouth, on the coal tanker *Broompark*. Frédéric remained in France with Irène, who was by now frequently indisposed, suffering from her long-term exposure to nuclear radiation.

Halban and Kowarski continued their research at Cambridge until in 1943 they and the heavy water moved once again, this time to the security of Canada. In the so-called Anglo-Canadian Collaboration they, and other French and British scientists, at last began the design of a nuclear reactor, moderated by heavy water.

THE RUSSIAN SECRET

In 1939, international phone calls were limited at best, so news in physics came by word of mouth or by letters and journals travelling across the oceans. In that time of slow-moving communications, some weeks

"Extremely Powerful Bombs"

elapsed before Bohr's paper of 1 September was widely known. In the Soviet Union, with war now taking place in mainland Europe, keeping abreast of events was especially difficult.

Russian physicists were already aware of fission because Igor Kurchatov had heard about it early from Frédéric Joliot-Curie. In the spring of 1940, Kurchatov pondered whether fast neutrons might induce fission in both uranium-235 and uranium-238 and initiate a chain reaction in natural uranium without the need for enrichment. He encouraged two junior colleagues—Georgii Flerov and Konstantin Petrjak—to use sources that emitted neutrons with different energies, put them inside a hollow uranium sphere, and then measure how the neutrons flowed. They found that fast neutrons create an insignificant amount of fission but also discovered something unexpected: fission occurs sporadically in uranium without any neutron bombardment at all.

They repeated the experiment underground in the Moscow Dinamo subway station to shield the apparatus from cosmic neutrons, which they suspected might be the cause of the phenomenon. However, *spontaneous* fission persisted. It was clearly real.

Spontaneous fission occurs because even natural uranium is slightly unstable. An explosive release of energy would require a chain reaction to build within a millisecond, and while spontaneous fission is too rare to be a source of energy, it can be enough to cause premature release of energy before a chain reaction takes over. Flerov and Petrjak reported their findings at the Soviet Academy of Sciences in May 1940. All the accumulating data seemed to mitigate against a nuclear explosion being possible. Kurchatov saw no reason for secrecy and a short report appeared in the American *Physical Review* in July that year.[7]

This attitude changed in the autumn of 1940 when two theorists in Saint Petersburg saw Bohr's paper and glimpsed its hidden threat. Bohr had observed that natural uranium contains so little U235 that the blanketing effects of the U238 will kill a rapid chain reaction. This raised an implicit question: what if there were more U235 in a sample? How much U235 would be needed to tip the scales? This was the question that Yulii Khariton and Yakov Zeldovich now addressed.

Khariton and Zeldovich were both Jews from different backgrounds. Zeldovich was Belarusian, age twenty-six, and working in Saint Petersburg. Khariton, age thirty-six, was more senior and world weary; born in Russia, he had moved with his divorced mother to Germany, but in 1928, horrified by the Nazis, they returned to Russia to settle in Saint Petersburg. Khariton and Zeldovich calculated that if uranium-235 was slightly higher in concentration—more like twenty rather than seven in every thousand atoms of uranium—a chain reaction could occur so long as the neutrons were moderated with water. Their calculations were correct except for one thing: they assumed normal water to be much more effective as a moderator than it is in practice.

The implication that a chain reaction and the possibility of an explosion might be possible were uranium to be enriched in its uranium-235 content was not news to advertise widely. Hitler and Stalin had signed a nonaggression pact, but that did not make them allies—as would be confirmed within a year when Germany invaded the Soviet Union. The Soviet Union classified this discovery secret. It remained so throughout the war and for many decades afterwards, only coming to light following the fall of the Berlin Wall. Unknown to the Russians, or indeed anyone outside a small handful of people at that time, was that in the United Kingdom two other Jewish émigrés had already found even starker results about the potential implications of enriching uranium-235.

"A RADIOACTIVE SUPERBOMB"

In the summer of 1939, Otto Frisch left Copenhagen to visit Birmingham University where Rudolf Peierls had settled. Frisch thought it would be a short visit, but as European countries were soon falling one by one to the Nazis, he stayed in Birmingham with his fellow émigré.

The two of them had read Bohr's paper and agreed that if fission were to be a practical energy source, uranium must first be enriched in U235. Frisch, the experimentalist, began some tests to see how feasible this might be. To do so, he used the phenomenon of thermal diffusion. When a mixture of two gases is in a container, one end of which is hot and the other cold, a gentle breeze develops in which the lighter gas molecules drift towards the hot end while the heavy ones accumulate at the cold end. Frisch used uranium hexafluoride, the only practical gaseous compound of uranium. He hoped to enrich the U235 by extracting gas from the hot end where the relatively light U235 would gather, then repeating the exercise over and over, which would progressively increase its concentration. The problem was it was very difficult, and he had no idea for how long he would have to continue to be successful. A tonne was out of the question, and the task looked impractical even if 100 kilogrammes were sufficient. Then in March 1940 he put the question to Peierls: how much U235 do I need to accumulate?

Peierls later recalled how nobody had really thought hard about how much separated uranium would be required to make an explosion, what is the *critical* size? Determining the critical size is analogous to answering the question: how big does a baby have to be before it can retain its body heat without swaddling? Body heat is produced within the child's volume whereas it is lost through the skin. Likewise, neutrons are produced by fission within the uranium and can escape through its surface.

Recall, when size increases overall, volumes grow in proportion to the cube of the length scale whereas surfaces increase as the length squared; the relative importance of surface to volume therefore declines inversely as the length increases. This means that for large volumes, such as a fully grown adult, or a large lump of uranium-235, the surface plays a relatively smaller role. In the case of uranium fission, the generation of neutrons taking place within is the analogue of heat production in a human: as a baby grows to become an adult, production of body heat internally and its surface loss reach equilibrium. The same principle holds for fission. The loss of neutrons through the surface becomes less important as the overall size of the uranium grows. The critical scale is that where the uranium lump is large enough that the production of new neutrons matches their loss through the surface. At sizes larger than this—becoming *supercritical*—the production exceeds the loss; conditions become ripe for an explosion.

So, how big is the critical amount for uranium-235? "To our surprise, if you worked this out on the back of an envelope, the amount came out to quite small", Peierls recalled.[8] Small indeed. Peierls's theoretical calculation suggested a terrifying reality: for the first time an atomic bomb that could destroy a city looked possible, requiring only an amount of uranium the size of a grapefruit.

Peierls recalled that they then asked themselves if you could make such an explosion, what would happen? "And again, on the back of another envelope, it came out that, while you couldn't predict the exact power, the effects would be enormous." He estimated that the time between successive steps in the chain reaction would be about five millionths of a second, the number of active neutrons doubling each time. After about eighty such generations, the lump of uranium would be vaporised into fragments smaller than the critical size, and fission would stop. By then, however, the damage would have been done.

The implication was staggering. In less than one thousandth of a second, the uranium would be hotter than the centre of the sun, and the explosive effects of this small piece of metal would be greater than several thousand tonnes of dynamite. Frisch recalled the moment: "We stared at one another and realised that a bomb might after all be possible". They were immediately struck by the worry that what was now known to them might already have been discovered by scientists in Germany. In fact, other than Frisch and Peierls in the United Kingdom, and Khariton and Zeldovich in Russia, no one elsewhere appears to have realised the potential of enriching uranium-235 and the practicality of this being a route to an atomic weapon. Peierls recalled: "We were elated in some sense but also frightened. The idea of Hitler getting this bomb first was certainly frightening. That was the main thing that made us 'step on it' and persuade other people to take it seriously."[9]

Such a weapon meant that a single plane would be able to deliver destruction that currently required an entire fleet of bombers. The enormity was that there would not be just a citywide explosive blast, but radiation would spread with the winds affecting larger areas, exposing hundreds of thousands of people. The sickness of Pierre and Marie Curie, and in Marie's case her death, showed that nuclear radiation's effects could develop over time. There was even the possibility of potential genetic modifications through the generations. This weapon appeared to be potentially destructive both instantaneously through its blast's spatial extent and in the effects of radiation continuing over decades.

In March 1940, Frisch and Peierls wrote a memorandum for the British government about the possibility of what they called a "radioactive super bomb". They pointed out that there is a critical size of a few kilogrammes, below which the material is absolutely safe. A quantity that exceeds the critical amount, however, is explosive. As for that explosion they noted "no material or structure could expect to resist the force of the explosion". While they hoped that no civilised nation would use such a device, they feared that the only protection against one would be to have one oneself.[10]

Sensitive to the needs for absolute secrecy, they typed the report themselves. They were working in old Nissen huts which were on the grounds of Birmingham University. It was a warm day, and the window was open. They heard a rustling noise outside, panicked, and went to have a look. They discovered that the "eavesdropper" was a lab technician innocently tending tomatoes along the south-facing wall of the building. They were much reassured that there wasn't anyone spying on them—at least not then.[11]

Frisch and Peierls's discovery gave the first frightening vision of nuclear reality. A government committee of scientists evaluated their report and immediately classified it top secret—so secret that even the two émigrés were excluded from the discussions. This nonsense was soon removed and the expertise of the bomb's fathers wisely consulted. Tests were made on the feasibility of enriching uranium-235 by diffusion of uranium hexafluoride gas through membranes, or by other methods such as centrifuges and thermal diffusion. Gaseous diffusion was identified as the most promising.[12]

TUBE ALLOYS

Frisch and Peierls were but two of the growing band of Jewish scientists fleeing from the Nazi scourge. It is ironic that Hitler's actions were providing the Allies with the very scientists who would help defeat the Axis powers. Among these was a Polish physicist named Joseph Rotblat.

Rotblat was born to a Jewish family in Warsaw in 1908. He studied physics at the University of Warsaw, gaining his doctorate in 1938, during which time he married a literature student, Tola Gryn. Following the discovery of fission, in early 1939 he conducted experiments at Warsaw and found that a large number of neutrons were emitted. He realised that as this happened in a very short time, it was possible that large amounts of energy could be released. From his measurements he concluded that generations of neutrons could accumulate in a millisecond and give rise to an explosion.

James Chadwick had left Cambridge in 1935 to become professor at Liverpool University, where he built an accelerator to study nuclear reactions. Rotblat wanted to build one in Warsaw, so he arranged to go to Liverpool to gain experience, but as he couldn't afford to take his wife he went alone, leaving her in Warsaw. Chadwick was quickly impressed by Rotblat's ability and arranged for him to be funded through a fellowship which doubled his salary. Now able to cover the cost of his wife, Rotblat returned to Warsaw intending that Tola come back with him. Unfortunately, she was ill after having had an appendix operation but decided to follow Joseph to England in a few days.

She who hesitates is lost. It was the end of August 1939, and before Tola could leave, war had broken out. Tola was trapped. Efforts to get her out via Denmark, using the influence of Niels Bohr, or via Italy were all closed off as those countries were soon overrun by the war.

Joseph never saw Tola again. She was a victim of the Holocaust and died in the Polish concentration camp at Belzec. These events profoundly affected him and, apart from a burning desire to eliminate the Nazis, he became a lifelong pacifist.

At Liverpool University, Rotblat and Chadwick, who had been made privy to Frisch and Peierls's proposal, made measurements of the interactions of neutrons, which were key to determining its practicability. Their experiments, together with investigations on the chain reaction by a team at the Cavendish Laboratory, all took place in total secrecy. After several months the British were confident that a realistic research and development project with the goal of constructing an atomic bomb was feasible.

Chadwick, discoverer of the neutron, former deputy to Rutherford, and since 1935 a Nobel laureate, was a highly respected member of the British scientific establishment. He wrote a formal report on the developments to Prime Minister Winston Churchill. Churchill agreed to proceed, thanks to his trusted science advisor Frederick Lindemann, Lord Cherwell, who urged him to go ahead with the attempt to create a British bomb. In August 1941, the British project began under the code name "Tube Alloys"—blandly chosen so were anyone to stumble across it by accident, they would think it to be no more than a study in metallurgy.

Even while writing the report, Chadwick was becoming convinced that an atomic bomb would be inevitable. He started taking sleeping pills due to thinking about this possibility that nuclear weapons could be real. For the rest of his life, he never stopped taking them.

CHURCHILL'S DOWRY

At the start of 1941, the United Kingdom was under threat of invasion and in terrible straits. The United States, meanwhile, was not yet on a war footing, nor was it clear whether they would want to become involved in what, for many of its citizens, was a European war. Winston Churchill was desperate to secure the support of President Roosevelt in what the British prime minister saw to be a war to preserve human decency. As part of the dowry that he was ready to offer to achieve that end, Churchill was prepared to share some of his nation's deepest scientific secrets, such as the development of radar and the work of its scientists showing that an atomic bomb made of uranium could be a viable proposition.

Unaware of these developments, physicists in America were proceeding with their own plans to harness nuclear energy and, possibly, produce an explosive device. Early in 1941, Ernest Lawrence in Berkeley converted his cyclotron into an electromagnetic separator—a means of using the cyclotron's magnetic fields to separate U235 from U238, thanks to their different masses. The principle is that when a beam of high-speed ions of U235 and U238 pass through a magnetic field, the lighter U235 is deflected more than the heavy U238. Lawrence had great hopes that this method of electromagnetic separation would prove an effective way of isolating U235.

There was no confidence, however, that uranium research would pay off in time, were the United States to go to war soon. By May 1941 this was the Americans' pessimistic outlook: not until 1943 would there be enough radioactive material to drop on an enemy; a nuclear reactor capable of powering naval vessels would take three or four years; the best that could be offered was a bomb of indeterminant power at some indefinite future, certainly not before 1945.

The idea of bomb production was focusing totally on *slow* neutrons—clear proof that much more needed to be done. There was certainly no expectation at this stage of any nuclear weapon being produced by the United States. Meanwhile in the United Kingdom there had been considerable progress building on Frisch and Peierls's original insight. Its scientists had established that if sufficiently pure uranium-235 were assembled, it would fission even with *fast* neutrons. Moreover, a critical mass of just 10 kilogrammes would be enough! A bomb that size could be loaded on an existing aircraft and be ready in two years. All of this was included in the report that Chadwick prepared in the summer of 1941.

The minutes of the British committee that reviewed this work was shared with the United States, as part of Churchill's desire to involve them in the war effort. These papers were received regularly by the chairman of the Uranium Committee, Lyman Briggs. Briggs, who had been head of the National Bureau of Standards, was a classic bureaucrat and generator of dull prose. Not only were his reports so bland that President Roosevelt would not have taken note of them, but obsessed by secrecy Briggs locked the British minutes in a safe without sharing the information with colleagues. American scientists were therefore completely ignorant of British progress.

Unaware of this folly, the British were concerned at the lack of feedback from the United States. In August 1941 Mark Oliphant visited the United States to find out what was happening. He visited Briggs in Washington, DC, and was apoplectic when he discovered that "this inarticulate and unimpressive man" had not shown the reports to members of his committee. Oliphant met Lawrence, Fermi, and other leading scientists, and was appalled to discover they knew nothing of the British advances. He then met with the Uranium Committee directly. He told its members they had to concentrate every effort on the bomb, not work on power plants. The bomb would cost twenty-five million dollars, but Britain didn't have the money or manpower, so it was up to the Americans. Members were surprised, Briggs having kept them in the dark. One of them said, "I thought we were making a power source for submarines".[13]

Having been primed by Oliphant, the American programme was ready when Chadwick's full report arrived. Once the US physicists were aware of the depth of British work, the Uranium Committee assessed that a bomb could be produced were there between 2 and 100 kilogrammes of uranium-235 available, and that to separate this amount of the isotope would cost between fifty and a hundred million dollars.[14]

Thanks to the insights of Frisch and Peierls, and fifteen months of work in the United Kingdom, the American programme to design and eventually construct an atomic weapon was now ready to move forwards. Any doubts that this would be a wartime effort were removed in December that year when the Japanese bombed Pearl Harbor, bringing America into the Second World War.

18
A Nuclear Engine

NEPTUNE AND PLUTO

Everyone had been focusing on the fission of uranium-235 and how to enrich uranium to produce pure samples of that key isotope. Then in 1940 another way to release nuclear energy explosively was found. This involved transuranic elements, the very things that Fermi back in 1934 had been chasing and erroneously thought that he had discovered.

The story begins soon after the world learned of fission in January 1939. Edwin McMillan, a thirty-two-year-old American physicist working at the cyclotron in Berkeley, was testing the idea of fission for himself by bombarding natural uranium with neutrons and looking for the fragments. These would fly through air, or other materials, for some distance, the range of a fragment depending on its mass. McMillan planned to identify the fission products by measuring their ranges. For his target he used powdered uranium oxide, which he spread on filter paper.

The experiments didn't reveal anything very interesting about the fragments, but he was surprised to discover that the filter paper next to the uranium had become radioactive. When he measured its intensity, he found there were two separate sources, one with a half-life of twenty-three minutes, the other of two days. These values differed from all known fission products. As this activity was in the vicinity of the uranium, and not blasted out like the fission fragments, he mused that it was radioactivity induced in the uranium itself by the neutrons.

Around 1935, Hahn and Meitner had discovered a radioactive source with a half-life of twenty-three minutes and identified it as "undoubtedly" uranium-239.[1] The other half-life of two days was something previously unknown. McMillan thought that this might be resulting from the uranium-239 undergoing beta decay within minutes and converting into element number 93, which decayed over days.

This was not sorted out until spring 1940 when McMillan was joined by Philip Abelson, a twenty-seven-year-old postdoctoral assistant. After a lot more irradiation to produce a bigger sample of the mystery product, Abelson investigated its chemical properties and concluded that they were indeed producing element number 93. He and McMillan submitted a paper to *Physical Review* with the title "Radioactive element 93". This was published in June 1940. They suggested in the paper that element 94 might also be made by a second beta decay taking place, namely the first one having converted uranium, element 92, into element 93, being followed by another converting 93 into 94.

In Cambridge, England, where research work was already classified secret, Swiss émigré Egon Bretscher had come to similar conclusions as the Americans. He was interested in what was happening when neutrons were captured by uranium-238, and the possibility that this could be the route to produce element 93 and even element 94. He and his colleague

A Nuclear Engine

Norman Feather—the same who had helped Chadwick confirm the discovery of the neutron back in 1932—used their understanding of nuclear stability to determine theoretically whether any of these new elements would be fissionable. Their calculations implied that number 94 would fission more easily even than uranium-235 when hit by either thermal or fast neutrons. This promised a considerable economy of raw material as it would be possible to use the fission not only of the minute fraction of U235 in natural uranium but also the more plentiful U238 as fuel to breed these transuranic elements. Bretscher was well versed in chemistry and realised that element 94 could be separated very easily from the bulk of uranium because it would be chemically different. Isolating element 94 would mean it would be easy to get pure fissionable material, so he immediately had a group of chemists work out how that could be done.

When doing his calculations in nuclear theory, Bretscher found also that U233 should be fissionable and that it could be created out of thorium in the same manner as element 94 was created out of U238. He knew that there was a lot of thorium on the west coast of India and realised that mining thorium could be a route to producing useable atomic energy via U233 for a long time. Weapons were necessary because of the war, but the excitement for him was this practical way of attaining the goal of unlimited energy.

One of Bretscher and Feather's colleagues in Cambridge was Nicholas Kemmer, who was born in Russia, grew up in Germany, and studied in Zurich. The polyglot wit spoke Russian, English, and German, was addicted to cryptic crossword puzzles, and revelled in puns. He proposed naming the new elements neptunium and plutonium, by analogy with the outer planets beyond Uranus, Neptune, and Pluto. His thinking was independently echoed in the United States where McMillan and Abelson had informally attached these very monikers to elements 93 and 94. Neptunium and plutonium are the names by which they are universally known today.[2]

The first experimental hints that these theoretical conjectures could be true came in 1941 when Berkeley physicist Glenn Seaborg confirmed neptunium undergoes beta decay to produce plutonium. By May 1941 he had shown also that plutonium is nearly twice as likely as uranium-235 to fission. The two countries were yet unaware of one another's interest in plutonium. Opinions about its suitability differed also.

To make plutonium in substantial amounts, one atom at a time, would require tonnes of uranium, intense neutron sources, and efficient means to moderate them. This would necessitate what we now know as a nuclear reactor. In Britain, with its focus immediately on the needs of the war and with research already well advanced in enriching uranium, the production of plutonium was regarded primarily as an opportunity for generating energy postwar. Churchill's science advisor, Lord Cherwell, dismissively said: "Who wants to win a war with an element no one's ever seen?"[3] In the United States, meanwhile, Enrico Fermi was aware that much of what physicists believed about the utility of enriched uranium and the production of plutonium was based on intuition and extrapolation of limited, small-scale, experimental investigations. To take the next step in releasing nuclear energy, he planned to build the first nuclear reactor and see if nature really matched what physicists anticipated in theory.[4]

THE NEW WORLD

In 1942, Enrico Fermi moved atomic power from theory towards reality. Back in 1939, as soon as the idea of a chain reaction in uranium was mooted, Fermi had established that graphite might slow the neutrons enough to achieve this and that uranium-235 was the key to causing fission. These experiments convinced him that a chain reaction could be maintained, at least in theory.

This is where things stood for three years: much of the understanding was "in theory". And although experiments showed that the fission of uranium liberates more than one neutron for each one lost, it was not a given that a chain reaction would occur in a large sample. Fermi's first attempts to build a reactor in 1941 at Columbia University showed that in practice the effective number of secondary neutrons produced on average by the initial neutron—a number we can denote by n—was only about 0.9. Recall, if the first impact produces on average n neutrons per fission, these will in turn produce n^2 at the second generation, the third generation will contain n^3, and so on. If n exceeds 1 in practice, the number of neutrons will grow exponentially; below 1, the reaction will die out. In other words, what Fermi had found was that although in the laboratory tests n was as big as 2, in his reactor this in practice turned out to be some 10 per cent less than required to maintain a chain reaction.

He understood why. In a reactor there is a lot of extraneous material which, while necessary in the construction, absorbs neutrons and reduces the overall efficiency. For example, iron canisters containing the uranium and impurities in the graphite all conspired to soak up precious neutrons in his first trial. After a year of experiments, with n getting larger but tantalisingly always short of the critical value of 1, Fermi believed he understood enough to design a reactor that could "go critical"—maintain a chain reaction at large scale.

So, early in 1942, in his new job at the University in Chicago, Fermi began a series of experiments geared towards his goal: a machine to demonstrate the reality of a self-sustaining chain reaction using neutrons and uranium. The previous year, Seaborg had shown that some uranium is converted ultimately to fissile plutonium, which inspired another application of Fermi's experiment: design of a much larger reactor to produce plutonium on an industrial scale.

One piece of theory on which everyone agreed was that in a small piece of uranium a chain reaction would die out as soon as neutrons exited from the surface. Another was that the neutrons emitted in the fission process were too fast to be effective in continuing the reaction. The goals for Fermi were to design a large reactor in which even the secondary neutrons are slowed, and to determine precisely how many neutrons on average are produced.

By November he was ready to make the key experiment. A team of scientists manhandled 6 tonnes of uranium metal and 50 tonnes of uranium oxide to provide the fission material, and 400 tonnes of graphite to moderate the neutrons. The graphite was in the form of bricks, each about the size of a backpack, which were assembled into a stack nearly 8 metres high and 6 metres wide, all located on a squash racquet court under the abandoned west stands of the university's football field.

The immense construction was mostly graphite, interspersed with pockets of uranium. This clever trick of mixing the graphite and uranium throughout the whole device meant that secondary—fast—neutrons produced by fission in the uranium had a good chance of passing immediately through more graphite and slowing. The reactor thereby produced generations of neutrons ideally tuned to induce further fissions and yet more neutrons. Fermi's team did this by alternating layers of pure graphite slabs with other layers in which uranium chunks were embedded in the graphite. It was this layered structure that gave rise to the description of the assembly as a nuclear *pile*.

Carefully engineered though the slabs were, graphite is dirty to handle. For them to fit flush to one another, they were first cut to 50-centimetre lengths and then smoothed by hand. Holes had to be drilled in them to hold the nuggets of uranium that would fuel the reactor. PhD scientists who climbed ladders to ferry the blocks—forty-five thousand in all, weighing about 8 kilogrammes each—quickly looked like coal miners emerging from a shift underground. Fermi's calculations implied that ideally the pile should have been 9 or 10 metres tall, but they were restricted by the height of the squash court. Nonetheless, if Fermi was right, the final pile ought to chain-react as long as it reached just below the ceiling.

Clever though Fermi was, there was no means of reliably accounting for possible shortfalls in the manufacture of the graphite. For example, impurities in the graphite might be converted into unwanted radioactive forms that poison the reactions and shut down the chain reaction. Fermi had made educated guesses at the likelihood of these negative effects and designed the pile to cover for them.

At the other extreme it was possible that everything might work more efficiently than anticipated, leading to a catastrophic release of nuclear energy and radioactive material into the Chicago atmosphere. To guard against this, he built in a safety mechanism in the form of metallic rods, made of cadmium. Cadmium is like a nuclear blotting paper that absorbs neutrons. Vertical slots in the pile contained three sets of cadmium rods, which absorbed so many neutrons that the pile rested inert. Fermi calculated that one rod would be sufficient to prevent a chain reaction, and as a safety mechanism in the event of an emergency, one of the rods was designed to fall under its own weight, back into place, killing the reaction.

Seventeen days after laying the first graphite block, the pile was complete and ready for the experiment to begin on the morning of 2 December 1942. Scientists arrived after trudging through the snow, a raw wind adding to the bitter cold. During the day, a small crowd of spectators accumulated, wrapped in coats and scarves as even in the dingy squash court the temperature was below freezing point. Of the small team actively involved in running the experiment, three—jokingly known as the "suicide squad"—were crouched on top of the pile, their heads touching the ceiling. If something catastrophic happened and the pile got out of control, their role would be to quench the reactor by flooding it with a cadmium solution.

The morning began with all but one of the rods being raised out of its housing. Fermi explained to the assembled watchers that the remaining rod would be raised at his instruction a little at a time. This would be done by their colleague, George Weil, who was on the floor operating a pulley attached to the top of the rod.

To utter silence, Fermi told Weil to "set the rod at 13 feet", meaning that 13 feet of the rod would still be inside the pile. Sensors designed to detect neutrons gave out audible clicks, and an automatic device recorded the intensity on a cylindrical roll of rotating graph paper.

Fermi performed calculations on his slide rule and verified that the results agreed with his predictions. Satisfied, he told Weil to raise the rod by a small amount. The counters clicked faster as the rod was raised, and then became steady at a new, higher, level. Fermi computed the result and again, to everyone's satisfaction—and some relief—it agreed with what he had expected.

They continued like this through the morning, Fermi issuing commands, Weil gently raising the rod, Fermi calculating, the ink marks on the graph paper making a line that moved ever upwards in a series of steps, one for each of Weil's manipulations of the rod. The audience could see that the curve on the diagram continuing to rise as the counters clicked ever faster. At last, with a dramatic pause Fermi announced that two or three more liftings of the rod should be enough to achieve a chain reaction generating energy of its own accord.

If at last they were about to achieve the goal of n larger than 1, the number of neutrons in successive generations would grow rapidly, ultimately exponentially without limit. An historic moment of great opportunity, but also of potential risk, beckoned. Now was the moment for a final check of his calculations. Instead of telling Weil to raise the rod again, Fermi turned to the audience and pronounced his familiar command: "Let's break for lunch!"

Fermi's calculations showed that n was still very slightly below one. The summit was in sight, but the approach to the mountain top was a long gentle slope. Each time Weil raised the rod, n increased nearer to one but by ever smaller steps. Fermi expected this. The mathematics implied that they would reach the threshold when Weil withdrew the rod one more foot.

Throughout the morning and into the early afternoon the results had agreed with what Fermi anticipated, giving everyone confidence that he was in control of the reactor, not the reverse. Even so, before taking the final step he decided to make one more safety check. He ordered that the safety rod be inserted. The neutron rate dropped immediately; the safety backup indeed worked. Confident that everything was now ready, Fermi told Weil to winch his rod up, one more foot.

Fermi turned to the gallery: "This should be enough!" he announced.

The clicks on the counters increased in intensity, but this time instead of settling, they were soon a continuous roar. The line on the graph also headed north and, instead of levelling out, kept on growing. The neutron intensity was already larger than the instruments could handle. Another set of instruments designed for greater intensities kicked in. One that could handle intensities ten to one hundred times greater recorded the growth, but this too was saturated as was soon that for one thousand.

The reaction had become self-sustaining. The neutrons were multiplying without limit, doubling in intensity every two minutes. With each fission, heat was being generated and there was no means to remove it—this was an experiment, not a machine for producing usable power. Nonetheless, in this moment was born a new era of practical production of nuclear energy, generating heat that could be supplied to reservoirs of water, turning them into steam which could then drive a turbine and produce electricity for the grid.

Had the Chicago experimental pile been left to run unchecked, it would have eventually melted. A meltdown such as would happen at Chernobyl could have occurred, but with the simplest of safety mechanisms such an eventuality is easily averted. Fermi was completely calm as he waited for four minutes and then ordered that the safety rod be inserted. Weil's rod and the other rods were all pushed back in and the clicking of the counters fell silent. Cadmium rods which absorbed the neutrons proved themselves as the on-off switches for the nuclear age, for it was at that moment, at 3:53 p.m. on 2 December 1942, that humans achieved the promethean dream of controlling the release of energy from the atomic nucleus. The overture heralding the nuclear age had begun.

A phone call sent the news of Fermi's success to the heart of the Uranium Committee in Washington. It was coded *en claire*: "The Italian navigator has just landed in the New World." "Were the natives friendly?" came the excited response. "Everyone landed safe and happy."[5]

THE ROAD TO TRINITY

Fermi's breakthrough of 2 December 1942 not only showed how to liberate nuclear power at a steady rate, with vast economic benefits for the future, but also demonstrated a means of creating plutonium. A nuclear reactor is a plutonium factory whose production line consists of neutrons multitasking. The fission of uranium provides power for the factory as well as liberates more neutrons to maintain the reactions. Some of those neutrons produce more nuclear fuel; others breed plutonium.

Recall, plutonium is highly unstable, any present in the Earth's infancy having long since decayed and vanished. Consequently, plutonium today is a synthetic element that must first be created—in the scientific jargon it must be *bred*—in a uranium reactor. So was born the twin application of uranium reactors: their advertised use as machines for liberating nuclear energy for the use of society, but also a way of breeding plutonium, one atom at a time, whose primary use would be as the fuel source for nuclear weapons.

In California at the Berkeley cyclotron—a particle accelerator by any other name—Ernest Lawrence had irradiated uranium with fast neutrons and successfully produced some plutonium, small quantities admittedly, but sufficient to enable its properties to be investigated. Experiments conducted by the Berkeley team soon showed that chain reactions in plutonium were indeed more likely than in uranium-235, making plutonium potentially a more efficient source of explosive nuclear energy.

Two routes to a nuclear weapon had thus come into play: enrichment of uranium and the breeding of plutonium. With war now raging,

A Nuclear Engine

work on nuclear reactors, and plutonium and uranium, was done in secret. Each of these would require a major industrial effort. The enterprise was focused on the United States, under the code name "Manhattan Engineer District", colloquially known as the Manhattan Project.

Meanwhile, British research into uranium—the Tube Alloys project—had made great progress. The British had focused on the theory of separating uranium isotopes by diffusion of gaseous uranium hexafluoride. The idea was for this gas to be in a cylinder with a piston at one end and a membrane at the other. The piston pushes the gas through the membrane. U235 being slightly lighter is forced through slightly faster, which makes the far side of the membrane richer in U235 than U238. Now repeat the process using that enriched sample. The first question facing the theorists was: how many repetitions are needed to achieve the required amount of enrichment? Other practical questions included: What are the optimum temperature and pressure to run the device? How big a facility would you need in practice to do this to make enough U235 for a weapon?

The answer to the latter question was: huge! To develop such a facility in the United Kingdom at that time was impractical due to the continued threat of invasion, and any major industrial plant would be ripe for attack by German aircraft. So, during 1943, plans were made for the leading British scientists to move over to the United States and add their expertise to the Manhattan Project. Although the Soviet Union was by that time an ally in the war against the Nazis, political suspicion of Stalin's real agenda led Winston Churchill and President Roosevelt to freeze the USSR out of the project.

Development of a weapon was ultimately the responsibility of the United States army overseen by the redoubtable Brigadier General Leslie Groves. Groves was a big man, nearly six feet tall, and very solid. His hips were invisible beneath an ample girth, Groves's trousers resisting gravity through friction with his rotund midriff. He was a dynamo, who had overseen construction of the Pentagon and knew the leading industrialists such as Union Carbide, Dupont, and Chrysler. The brusque and militarily efficient Groves created as short and punchy a mission statement as there has ever been: "To get a bomb and be able to deliver it before the enemy can."[6]

He was also an astute judge of people. Groves, who was wary of academic Nobel laureates, made an unlikely choice as scientific director of it all: J. Robert Oppenheimer. Oppenheimer had made significant contributions to the theory of quantum mechanics but nothing so exceptional as to put him in contention for Nobel Prizes. He was charismatic, highly ambitious, and frustrated that his work had not brought him the recognition he felt he deserved. He quickly realised that the role of scientific director of the Manhattan Project could be his path to immortality. Three years after he had drawn an outline cartoon of a bomb on his office blackboard in Berkeley, Oppenheimer now led an international army of brainpower charged with successfully delivering an atomic weapon.

The difficulty of releasing nuclear energy in an explosive blast was all too apparent. The collective intelligence of physicists from across the continental United States and the United Kingdom, including many Jewish émigrés from mainland Europe, focused on the task. Graduate students and recent postdocs joined distinguished famous scientists. The average age was twenty-four, the most famous of these youngsters being Richard Feynman, already renowned as the new wunderkind.[7] After the war, and still not thirty years old, Feynman would revolutionise the understanding of quantum mechanics and himself become a Nobel laureate. During the Manhattan Project, Feynman's great attribute was a fearless ability to ask penetrating questions of the most distinguished scientists, and then improve on their answers. The challenge was as much one of engineering and industry as of science. Niels Bohr presciently observed that to build an atomic bomb "you would need to turn the entire country into a factory".[8]

From April 1943, the nerve centre of the Manhattan Project where the bomb would eventually be assembled was at Los Alamos in New Mexico, on a mesa 2200 metres above sea level. Remote and uninhabited, Los Alamos was the perfect spot to do the unimaginable. What had previously been wilderness changed completely. Laboratories for experiments, cabins for the scientists and their families, barracks for the military personnel, and offices for administrative staff were built in no time at all. The British may have invented the bomb, but the Americans were going to build it.

A nationwide network of collaborations between government, US companies, and academia was essential. Within weeks of Groves's appointment, a vast industrial enterprise began to spread its tentacles across the continental United States. Hundreds of manufacturers in nearly every state produced components; work was done in universities on the design of reactors, the production of plutonium, the enrichment of uranium, and the chemistry and metallurgy of these elements. Three main sites were run by the military, fuel for the bomb being prepared at two sites—Oak Ridge in Tennessee where uranium-235 was separated, and Hanford in the Pacific Northwest where plutonium was synthesised—and the final assembly of the weapons taking place at Los Alamos in New Mexico. More than one hundred thousand people were employed, ranging from labourers and technicians to Nobel Prize winners. The scale and investment were about equivalent to the US auto industry at the time.[9]

Work on separating uranium through centrifuges together with critical measurements on the properties of uranium and plutonium took place in Berkeley. The metallurgy of these toxic elements and how to handle them safely were investigated in Chicago. Although today centrifuges are the standard way of enriching uranium, in 1943 the technology was still primitive and gaseous diffusion—the original path that the British team had developed—was the chosen route. This is very difficult, however, more so than isolating plutonium, for example. Uranium-235 must be separated from other isotopes of the same element, which is done by focusing on their marginally different masses. To differentiate this mere one part in eighty and separate the components requires precision. Whether done by diffusion, centrifuge, or other means, it involves large engineering projects operating over long timescales. The plant at Oak Ridge contained a cascade of several thousand diffusion tanks, the largest over 1000 gallons (3800 litres) in volume, all contained in a U-shaped building, each leg of which was half a mile long. Enrichment of uranium to a level required for a bomb took several months. This contrasts the isolation of plutonium, which once synthesised is a different element and can be straightforwardly separated by chemical means.

However, first plutonium must be made. The small traces of the element produced at Berkeley had enabled its properties to be determined, but to synthesise enough for a bomb, large-scale nuclear reactors would be required. Fermi's successful experiments with the reactor in Chicago showed the way. Within twelve days of his demonstration, searches were underway for a production site.

A pilot reactor at Oak Ridge produced a few grammes of plutonium, but the factory needed for a weapon would have to be on a different scale. The site at Oak Ridge was deemed inappropriate for this, for several reasons. First, it was but 20 miles from Knoxville, a large town in Tennessee, and the possibility of a catastrophic accident at the site could lead to many deaths and severe health problems. Also, even a minor mishap might interfere with the uranium separation activities at the laboratory.

The decision was made to find a remote site that was about 200 square miles in area having no towns within 20 miles of its outer boundary. For safety reasons, the laboratories on the site would have to be at least 8 miles away from the reactors and from the plant for separating the newly synthesised plutonium. To cool the reactors, 100,000 litres of water would be required every minute, or 1600 litres every second. Above all, the operation would need a huge electricity supply.

The chosen location was in Hanford, a remote area in the southeast of Washington state, near the Columbia River. A high-voltage power line from the Grand Coulee Dam ran through the site, and there was also an electrical substation on its edge. The area was covered in sagebrush, where sheep grazed, along with a substantial amount of farmland. Given the needs of the war, Groves decreed that work should begin only after the summer harvest was done. So not until August 1943 did construction start. Some 45,000 workers built the reactor, separation plant, and distant laboratories. The first reactor was completed in September 1944 and produced measurable amounts of plutonium by November.

The design was inspired by Fermi's reactor, which used graphite as moderators intermingled with cylinders of uranium fuel, augmented by the essential addition of water to prevent the device overheating and igniting. Twenty thousand uranium slugs, each about 4 centimetres in diameter and 20 centimetres long, were needed to get the reactor started, and another two thousand were required every month to replace those removed for plutonium extraction. The extraction plant was 10 miles away, and a railway line linked it to the reactor. Everything was handled by remote control. Items were moved by an overhead crane as once the uranium slugs were irradiated, they themselves became highly radioactive and a hazard for humans. Closed-circuit television monitored what was happening in the reactor and in the separation plant. Nuclear science became automated.

Although plutonium is easy to separate, it is very nasty to work with. It is highly radioactive, toxically poisonous like heavy metals, and fissions spontaneously. Its metallurgy is complex, which created a problem: how does one prepare plutonium in a form suitable for a bomb?

As carbon comes in different allotropic forms—diamond, graphite, graphene—so does plutonium. One form, known as alpha, is like chalk, which would be unsuitable for a weapon. Another form, known as delta, is metallic. This is ideal except that it is unstable and morphs into the alpha form. The plutonium harvested from a reactor must be stabilised as an alloy if it is to be of any use. The metallurgists found that mixing it with gallium—a soft silvery metal like aluminium—does that. This was a key discovery, but for which the nuclear potential of plutonium might have been worthless.

For more than two years research and development in the many satellites of the Manhattan Project combined with an intensive series of experiments and theoretical investigations at Los Alamos itself. The conception of an atomic bomb in the minds of theorists had been straightforward, but its gestation proved difficult; at times it seemed that, at least in the case of the plutonium weapon, the project might fail.

Recall that for an atomic bomb to work, there needs to be enough fissile material (uranium-235 or plutonium) present simultaneously to make a chain reaction explosive and then to keep the process going. This is the logic behind the concept of critical size, below which fission cannot continue, rendering a sample harmless. For this reason, an atomic bomb must be assembled from separate pieces, each of which is less than the critical amount. Experiments at Los Alamos and Berkeley had established that for uranium-235 and plutonium-239, these are about the sizes of a soccer ball and a grapefruit, respectively—in mass, for a solid sphere, some 50 kilogrammes and 10 kilogrammes. For uranium, this is larger than Frisch and Peierls's original back-of-the-envelope calculation, but it is still small. These harmless subcritical pieces are brought together making a supercritical, explosive amount.

For uranium, which was reasonably stable, this engineering was straightforward: one subcritical piece would be fired from a gun into a static piece of uranium and the chain reaction would exponentiate. Experiments by Emilio Segrè in 1943 had measured the chance that uranium fissions spontaneously and found it to be sufficiently low that the subcritical pieces of uranium would remain pure enough to create an explosive mix. However, he discovered that for plutonium spontaneous fission is likely to initiate the chain reaction prematurely, before the individual pieces came together. The implication: a gun mechanism such as used in uranium would lead to the plutonium pre-detonating—a sputter, not an explosion. This was utterly unexpected. A more powerful gun capable of ramming the plutonium pieces together at 1000 metres per second, rapidly enough to beat the degradation from spontaneous fission, would melt the plutonium bullet and target before they had time to join. By April 1944 a practical plutonium bomb seemed impossible.

THE DEMIGOD

In his 1973 BBC TV series *The Ascent of Man*, Jacob Bronowski called John von Neumann "the cleverest man I've ever known". Among his many talents, von Neumann could recite vast tracts of text after just a single reading and perform mental arithmetic at lightning speed. At Princeton in the 1930s his brilliance was explained: von Neumann is a demigod who has come to earth, observed human behaviour, and imitated it so perfectly that he appears to be one of us.

Bronowski added, "He was a genius, in the sense that a genius has two great ideas". Bronowski here was referring to von Neumann's development of game theory and its application to economics, and his singular role in developing computer science, both of which founded new fields of intellectual endeavour. Scientists who had been with von Neumann at Los Alamos knew of his other great ideas, but these were still classified secret in 1973 when Bronowski made his paean.

Von Neumann was born in Budapest in 1903, the son of a Jewish lawyer. That he was a mathematical prodigy was soon apparent, and in 1933 he became one of the first faculty members at the newly created Institute for Advanced Study in Princeton, New Jersey. With the Nazis' growing power in Europe, von Neumann remained in the United States for the rest of his career. He was no nerd. Indeed, the contrary, as he was a very social person who loved parties, high living, and bawdy limericks. His colleague at the Institute, Albert Einstein, perhaps threatened by von Neumann's genius, belittled him with the Germanic description *denktier*—a thinking animal. Von Neumann's skills were called on at Los Alamos in 1944 to confront the apparently insuperable problem of how to make a plutonium bomb explode.

A way of assembling the critical mass faster than a gun mechanism was needed. The scientists assessed that implosion, using chemical explosives to squeeze a hollow shell of plutonium segments into a small solid ball, could form the critical mass faster than any gun. Unfortunately, attempts to make detonation waves converge to a common centre failed as the waves from the various explosive charges merged and produced a mixture of high- and low-pressure regions, like eddies and splashes when streams of water merge.

The Los Alamos talent included several experts in conventional high explosives. One of these was a tall, languid Englishman named James Tuck. It would be Tuck who made the first inroad to a solution.

In the United Kingdom, Tuck had designed shaped charges, high explosives arranged in a hollow cone so that the resultant blast concentrates to its tip. This arrangement had proved effective at bursting through tank armour, and now at Los Alamos he suggested the technique might be adapted and applied to the plutonium problem. His point was that rather than trying to smooth out the peaks and troughs of the pressure waves, they should arrange the explosives to produce a converging wave from the start.

Tuck had found the way. How to do this in practice was a problem for the mathematicians to solve.

Hans Bethe, who was head of the theoretical division at Los Alamos, brought Rudolf Peierls to the laboratory to help develop the theory on how to do this. Peierls was joined by his colleague Klaus Fuchs and von Neumann. How to make a perfectly symmetrical implosion was compared to the challenge of compressing a glass of beer with an air gun without blowing any of the froth out of it. Von Neumann is credited with the idea of how to achieve the necessary amount of symmetry. He proposed that a "lens" of slower and faster burning explosive charges be arranged at the corners of hexagons and pentagons on the surface of a hollow sphere of plutonium, a pattern later made famous in the 1970s construction of soccer balls. The chemical high explosives then would force the pieces of plutonium inwards, their combined shock waves squeezing the shell rapidly into a solid dense ball. Von Neumann had turned Tuck's idea into a realistic way forwards—in theory at least.

The plutonium factory at Hanford produced its first batch of refined plutonium at the start of 1945 and delivered it to Los Alamos on 5 February. By this time there were three identical reactors completed on the site. By 8 March, they were running at full power, and by April kilogramme quantities of plutonium were being shipped to Los Alamos. By July there was enough ready for the experimental test—Trinity—and, if successful, for two further bombs for military use.

Even up to the moments immediately before the bomb was detonated, there remained one uncertainty: would much of the energy be released before dispersal of the fissile material brought the chain reaction to an end? In short: would a bomb explode or just liberate energy in a fizzle? All the theory implied it should explode, but ultimately only an experimental test could provide the proof. The answer to this final question came with the test of the plutonium weapon, code named "Trinity", at dawn on 16 July 1945.

This, the moment when the awesome energy of the atomic nucleus was released unconstrained for the first time, was when the nuclear age fully began. Oppenheimer famously commented: "Now I am become Death, the Destroyer of Worlds".[10] Another watcher exclaimed "Now we are all the sons of bitches", which more prosaically described what the scientists' achievement would make them. Only following Trinity did the enormity of the venture hit home for many. A few weeks later, it was in Japan that the first explosions of a uranium weapon took place—over Hiroshima on 6 August 1945—and the reality of Frisch and Peierls's apocalyptic vision was demonstrated to the world. A plutonium bomb exploded over Nagasaki three days later.

Hiroshima was a sprawling metropolis of a quarter of a million people spread over the marshy flatlands of the Ota River delta and surrounded by hills. The blast spread unhindered over the plains, causing complete destruction over two miles wide and killing one third of Hiroshima's 340,000 residents. The firestorm injured almost as many, some of whom would suffer slow and painful deaths over several years, caused by the bomb's nuclear radiation. The city of Nagasaki is set among hill ranges, granting some small protection to those living in the shadow of the blast. Even so, 40,000 people out of a population of 200,000 were killed outright, and 25,000 more were seriously injured.[11]

The numbers were huge in part because in all the calculations one thing had been overlooked. The military had assumed casualties would be limited because people in air raid shelters would be protected from the lethal gamma rays. But the shelters were not used: no "air raid" took place. The firebombing of Dresden and Tokyo, for example, had involved hundreds of planes, bombers, and fighters in support, and everyone had sought cover. At Hiroshima and again at Nagasaki, there were but three planes—the bomber, an observation plane with scientific instruments, and a camera plane—so no one realised what was about to ensue. At Hiroshima, at 8:15 a.m. people in the outdoors, including children on the way to school, were fried by the radiation. Days later, at Nagasaki at 11:00 a.m. on a warm summer morning, the disaster was repeated.

Few at Los Alamos would deny that what had attracted them had been the feeling of excitement at the science involved. The dynamics of a chain reaction, of matter at temperatures and pressures akin to those in the stars, all resulting in a colossal release of energy were deep mysteries whose solutions would take scientific understanding to new frontiers. Some, however, had understood the awful reality already and refused to take part or to complete the course. Lise Meitner, for example, had been invited to join the Manhattan Project but refused: "I will have nothing to do with a bomb"; she was deeply offended when a newspaper report of the explosions over Japan described her as the "mother of the atomic bomb".[12] Joseph Rotblat was the only scientist to quit the project when it became clear, by the end of 1944, that an atomic weapon to defeat Germany was no longer needed. And on 17 July, the day after the Trinity test, seventy scientists led by Leo Szilard petitioned the US president not to use the bomb over Japan, but to no avail.[13]

The full power of the nucleus had been released; you cannot undiscover what has been discovered. Fritsch and Peierls's 1940 prediction that no material could resist the blast and radiation was tragically confirmed. Their further observation that the only defence against such weapons would be to have them oneself presciently foresaw the reactions of world governments who for more than seventy years have followed a faith of mutually assured destruction.

SECRET NO MORE

The Allied progress in developing nuclear weapons had been passed to the Soviet Union by spies from the very start. Even Frisch and Peierls's top-secret memorandum of 1940 was known in Moscow within weeks due to John Cairncross, a spy in the British Civil Service. At the heart of the project itself, physicist Klaus Fuchs passed complete information to the Soviet Union for the entirety of the Manhattan Project and beyond, from the moment of joining the British team in 1941 until his arrest in 1950.[14] While intelligence from Fuchs and other spies proved invaluable to the Russians, the idea that the project itself could have remained secret, as Groves believed, proved fanciful. The Russians had already deduced what was going on simply from checking the open physics literature.

Before the Manhattan Project, before Tube Alloys, even before the breakthrough by Frisch and Peierls which began all the secrecy, physicists around the world had been absorbed by the implications of nuclear fission. In Russia, recall how Igor Kurchatov had realised its significance, which led two of his assistants to do their celebrated experiments in the Moscow subway. One of them, Georgii Flerov, was by 1942 in the Russian army, fighting at the front near Voronezh, about 300 miles south of Moscow. A brief weekend of leave by chance would convert him into an intelligence analyst.

Having a few hours to spare, he went to the local university physics department. Voronezh University had been evacuated, but the library shelves were full of the latest international journals. Keen to learn what progress had been made with nuclear fission, he looked through the pages of the available Western physics journals. He found papers on a variety of topics, but surprisingly nothing at all on the subject that fascinated him, which two years earlier had been the most sensational discovery in atomic physics since that of the nucleus itself. That was but half of it: not only had all mention of fission disappeared, but the leading nuclear physicists had also—Enrico Fermi, Niels Bohr, and other luminaries had published nothing for several months. The explanation for the absence of papers, Flerov quickly realised, was that American research on fission had become secret. This also explained the disappearance of the nuclear scientists: they were working on a nuclear weapon.

Flerov mused that at least the Americans and Soviets were on the same side in the war, but he worried that German scientists might have come to the same conclusion. He sounded the alarm by writing to Stalin in April 1942. What he didn't know was that Stalin was already aware of the state of the Allied programme, thanks to Fuchs, and that Flerov's former mentor, Igor Kurchatov, would soon be leading a secret programme to develop a Soviet bomb along those very lines. In mid-July 1942, Flerov was recalled from the southwestern front, summoned to Moscow, and put back into neutron research. During October, while the Battle of Stalingrad was at its height, Kurchatov was put in charge of the Soviet Union's quest for an atomic bomb.

After the war the knowledge of how to make atomic bombs proliferated, and the USSR exploded its first such weapon in August 1949. However, atomic bombs turned out to be just preludes to an even more terrifying weapon that could literally destroy worlds: the fusion or hydrogen bomb. For within days of the Trinity test having proved the reality of a nuclear explosion, and before the horrors of atomic weapons had been demonstrated at Hiroshima, Fermi was giving secret lectures to the scientists at Los Alamos. As he explained, in the realm of nuclear energy, atomic bombs are mere trifles.

19
Destroyers of Worlds

AN EXPLODING SUN

Terrible though atomic bombs are, their power is limited due to the need for a supercritical mass to be available. When that atomic bomb explodes, the material disassembles, so even if a huge amount of uranium or plutonium is put together, the active explosion destroys the mix and stops the chain reaction. The bombs that exploded over Japan were about 3 metres across, weighed roughly 4 tonnes, and had an explosive power equivalent to about 20 kilotons of TNT. The maximum in theory is about a megaton. Thermonuclear weapons—colloquially known as hydrogen or H-bombs—are theoretically boundless.

Detonation of one H-bomb, if it were big enough, could destroy all life on Earth.

They're called *thermonuclear* because they need the heat of the stars to ignite them. The sun is made of hydrogen, and it is fusion—the coming together and merging of protons, hydrogen nuclei, in its heart—that produces solar energy. Similarly, the fusion of hydrogen nuclei is the basis of thermonuclear weapons: hence hydrogen bombs, or fusion weapons.

A hydrogen bomb in effect brings the power of the sun to Earth. Although atom for atom the energy released in fission is about ten times larger than that in a hydrogen fusion reaction, the mass of these hydrogen isotopes is only about a fiftieth of the mass of uranium or plutonium. So, every kilogramme of a hydrogen bomb contains fifty times as many atoms as in the equivalent weight of an atomic—fission—bomb. The result is that you get about five times as much energy per kilogramme in an H-bomb than in an atomic fission bomb because there are more atoms involved. For example, as the fission of a kilogramme of uranium could produce a blast equivalent to about 20 kilotons of TNT, the fusion of a similar weight of deuterium and tritium would produce a blast equivalent to some 100 kilotons.

The story of the hydrogen bomb began back in September 1941 at Columbia University in New York. Enrico Fermi was by then a professor there, and one of his colleagues, Edward Teller, was working with the Uranium Committee exploring the possibility of nuclear weapons.

Teller was both a brilliant physicist and a talented pianist but cursed with a volatile personality. Born in 1908, he was seven years younger than Fermi. The political climate in Hungary during his youth and his experiences as a student in Germany instilled a lifelong hatred of both communism and fascism. One day in Munich, aged twenty, he had jumped off a tram while it was still moving and fell, and one of its wheels nearly severed his right foot. For the rest of his life, he walked with a limp. Self-obsessed and argumentative, he held strong opinions from which he was not easily swayed and would describe his ideas overconfidently. Difficulties with interpersonal relationships were a leitmotif throughout his life.

Fermi and Teller regularly had lunch together in the faculty club, and one day on their way back to the physics department Fermi stopped and posed the question: if the nuclear bomb works, "couldn't such an explosion be used to start something similar to the reactions in the sun?" He added that the device would not use hydrogen but deuterium for which the probability of the reaction happening is very much greater.[1] What Fermi had realised was, first, on Earth hydrogen nuclei are normally at atoms' length and do not fuse, but at very high temperatures—temperatures like those which occur in the sun's heart—the kinetic energy is sufficient to overcome the protons' mutual electrical repulsion. This enables fusion and the release of energy. His key insight was then that in an atomic explosion the temperatures would be millions of degrees: an atomic explosion can itself be the spark to ignite a thermonuclear explosion. Teller thought about this for some months, and although he could not develop a viable model of how to do this, nor could he convince himself that Fermi was wrong.

The next development in this story came the following summer, in 1942, on a train journey to San Francisco. Teller shared a compartment with Hans Bethe en route to a meeting that Oppenheimer had organised to study the design of atomic weapons. Bethe was a professor at Cornell in upstate New York while Teller was at Chicago, so Bethe had stopped off in Chicago both to pick up Teller and to see how Fermi's reactor work was progressing. Bethe was convinced that the reactor would operate and that a nuclear weapon would also probably work.

In their private compartment on the train, they could talk freely. Teller told Bethe about the idea of using the reactor to make plutonium, and then using the plutonium as fuel for a nuclear weapon. Teller, who always liked to jump to conclusions, cavalierly assured Bethe that the atomic bomb was probably a sure thing and proposed they should really be thinking about using the high temperatures released by an atomic bomb to ignite deuterium in a hydrogen bomb. Teller proved persuasive, later recalling that "by the time we reached Berkeley, Hans was equally fascinated with the possibility of a thermonuclear reaction".[2]

Bethe recalled that most of the time that summer was spent thinking about the possibility of a hydrogen super weapon. They encountered one problem after another and in many cases came up with solutions, but the difficulties always outnumbered the ideas. Meanwhile, Bethe began to ask himself why he was working on this. It was obvious to him that the atomic bomb had to be made because the Germans were probably doing so already, and that motivated him, but what Teller now described as the "Super Bomb" was a terrible thing, which if it worked could kill millions of people at random. Less a weapon of war than one of murder; Bethe decided that were a hydrogen bomb to be developed, he didn't want to be involved.

The threat—or opportunity—of a hydrogen bomb is that the more hydrogen you start with, the more fusions can occur. There is no disassembly or critical size constraint as in the case of fission. The only limit is a practical one of the amount of fuel that can be brought together, ignited, and then kept hot enough for fusion to continue. The power of a fusion bomb is therefore in earthly terms boundless.

The proof is blindingly obvious, quite literally: the sun is a vast fusion engine. This was what Fermi was alluding to in his remarks to Teller. The sun burns very slowly, however, which is what inspired Fermi's remarks about using deuterium. The fusion of protons in the sun requires a simultaneous beta radioactivity to occur, to convert a proton into a neutron and enable the reaction to proceed. This is so rare that after five billion years, the sun has only used up half of its hydrogen—in other words, were you a proton in the sun, there is still only a fifty-fifty chance that you would have been fuel for its fusion engine. By using deuterium, which already contains a neutron, Fermi hoped to speed up these reactions so that they could happen within a fraction of a second and liberate solar energy explosively. However, this proved to be easier said than done. Teller could not see how to, but the idea was so compelling that he was determined to find a solution and would do so with messianic zeal.

PERPETUAL MOTION

If the original vision had been that the spark of an atomic explosion would ignite the deuterium fuel like a burning taper being thrown into a can of gasoline, theoretical analysis soon showed that the reality would be nearer that of a match trying to light a log fire without any kindling. A small piece of the log might briefly glow, but there would not be enough heat to spread fire throughout it.

This was the main and persistent problem that Bethe and Teller were confronted with the moment they started their analysis that summer. The heat generated by a fission explosion would rip electrons from the atoms of whatever was around. When these electrons interacted with atomic nuclei in the deuterium fuel, they would be accelerated. When electric charges suddenly alter their motion, they radiate light. In this case the radiation leaks out and cools the deuterium. In short, while the atomic explosion would briefly light one end of the fuel, the ignition would not propagate.

Teller, however, was obsessed with the idea and convinced that he could find a solution. He came up with one scheme after another, but none of them worked. Isidor Rabi, another of the Jewish émigrés who became part of the Manhattan Project, said that Teller reminded him of another man who used to visit Rabi regularly to tell him about a perpetual motion machine that he had supposedly invented. Perpetual motion is not possible, and Rabi would explain carefully why that idea would not work. The man would then thank him and go away, only to return in a few days with a new design of a perpetual motion machine. This is how Rabi remembers Teller's obsession with the "Super" throughout the war.

While the primary goal of the work at Los Alamos was to design a fission bomb, Teller persistently berated Oppenheimer for not putting a dedicated team to work on a design for the Super. When Oppenheimer explained the needs of war compelled a pragmatic focus on the atomic bomb, not the Super—for which the trigger needed a successful atomic bomb anyway—Teller threw a tantrum. Oppenheimer resolved the face-off by allowing Teller to investigate the hydrogen bomb for a post-war programme, at the price of relinquishing his formal role in the Los Alamos theory group.[3] The result was Teller repetitively announcing he had solved the problem only to have its flaws exposed.

Fermi had directed his efforts to development of the atomic bomb. Following the success of the Trinity test, he looked again into the question of a hydrogen bomb. Over several months in the latter half of 1945, he gave a series of six lectures to the closed group at Los Alamos with his analysis of what was known as the Classical Super bomb.

Fermi discussed how fusion is initiated in the Classical Super, how energy is lost, and his ideas on how to combat this. Fusion involves isotopes of hydrogen. While the nucleus of ordinary hydrogen contains just a proton, that of the deuteron consists of one proton and one neutron, while the nucleus of tritium—the triton—consists of one proton and two neutrons. These are respectively denoted p (proton), d (deuteron), and t (triton).

The basic reaction that interested Fermi is two deuterons fusing and making a triton and a proton. This preserves the total number of nucleons and individual protons but rearranges them (Figure 9). This liberates energy because the sum of the masses of the triton and proton is less than that of two deuterons, and by Einstein's equivalence of mass and energy, this difference is manifested mainly as kinetic energy of the proton. The reaction only takes place when the two deuterons come into contact—literally fuse. To do so they must first overcome their mutual electrical repulsion. In his first lecture, Fermi proved that a temperature of 1,000,000 degrees Celsius from the atomic explosion is enough to do this. The fusion of the two deuterons generates heat and the temperature rises. The fusion reaction has begun.

So far, so good, as we appear to be en route to a self-sustaining reaction. But now the analogy of the log comes into play, and the problem

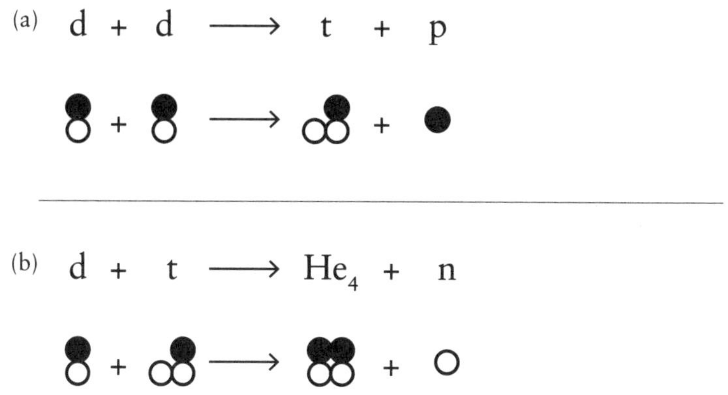

(a) d + d ⟶ t + p

(b) d + t ⟶ He₄ + n

Figure 9. (a) Two deuterons convert to a triton and a proton. (b) A deuteron and a triton produce an alpha particle (He$_4$) and a neutron. Protons are shown as solid circles and neutrons as open circles.

that Bethe and Teller had met returned: most of the deuterium fuel is still remote from the ignition point. Near this hot spot, energy is lost in the form of electromagnetic radiation—the deuterium cools. To have enough heat to set it all alight, a temperature of about 300 million degrees Celsius would be needed, Fermi declared, a huge challenge to achieve. How could some atomic kindling help set the whole deuterium "log" alight?

The following week, one day after the atomic explosion over Hiroshima, Fermi extended his analysis to include the effect of secondary reactions. The fusion of two deuterons produced some tritium (Figure 9a), which opened the possibility that this newborn tritium could in turn fuse with some of the deuterium, as in Figure 9b. Tritium is a very good nuclear fuel, and fusion of deuterium with tritium ignites at lower temperatures than that of pure deuterium, making ignition easier. Fermi calculated that this tritium would reduce the ignition temperature to 200 million degrees—lower than with deuterium alone, but still inconveniently high. He went away to calculate how much tritium would be needed to get the ignition temperature low enough to be within practical reach.

He gave his answer on 18 August, three days after Japan's surrender that ended the Second World War. He calculated that if one part in two hundred of the fuel consists of tritium, the ignition temperature would fall to 100 million degrees, which looked more promising. However, tritium is nasty. Unlike deuterium, which is stable and available, tritium is radioactive, difficult to handle, and expensive to manufacture. As a measure of the problem, Fermi noted that only a few thimblefuls of tritium gas had been made at Oak Ridge National Laboratory by a beam of neutrons hitting an isotope of lithium, lithium-6. During the next year, calculations confirmed that the amount of tritium for this Classical Super were impractically large.

Even were some faster ways of producing tritium to be found, Fermi identified a new problem. The time taken to reach the ignition temperature would be hundreds of microseconds, which while short to our senses is long on the timescales of nuclear reactions.

In 1940, Peierls had estimated that the interval between successive fissions could be a few microseconds. Subsequent measurements and improved theoretical modelling showed the time to be much shorter, for fast neutrons being nearer 10 nanoseconds or one hundredth of a microsecond. The mathematical models of many nuclear processes involved this sort of timescale, and so the Manhattan Project physicists duly invented a new unit of measurement—the shake. One shake equates to 10 nanoseconds and is typically the average time it takes for a newborn fission neutron to find another nucleus and, by metaphorically shaking hands with it, initiate the next step in a chain reaction. As nuclear clocks operate at the scale of shakes, Fermi's audience immediately understood that one hundred microseconds, the time he estimated it would take to ignite the Super, corresponds to ten thousand shakes, which is huge. Fermi concluded that inertia would not hold the system together that long.

On 11 September, in his fourth lecture, he investigated the question of cooling due to the radiation of photons from the electrons within the plasma of deuterium. This turned out to be large. The challenge now became how to reduce this, and by the fifth lecture, a week later, Fermi suggested that a magnetic field could hold the electrons away from the walls and reduce the loss of energy. This idea would prove to be relevant to modern fusion tokamaks—such as the Joint European Torus in England—whose inability to liberate fusion energy in useful amounts, eight decades after Fermi's lectures, highlights the difficulties that he was confronting back then.

In his final lecture, in October, he tried to find loopholes, but his conclusion was clear: the Classical Super would not in practice work. He ended with a wry dig at Teller, whose optimistic proselytising for the Super had become a source of amusement in Los Alamos. Fermi used his own nickname—Il Papa, the Pope—in his remark. He said: "Teller, who has overseen most of the work reported, is inclined to be more optimistic than is the lecturer. The procedure that has been adopted in trying to resolve the practicability of the Super is that Teller shall propose a tentative design which he considers somewhat over designed, and the lecturer will try to show that it is under designed." This brought Fermi to his punchline: "This makes the Pope the Devil's advocate!"[4]

VON NEUMANN'S GREAT IDEAS

One of what Jacob Bronowski lauded as von Neumann's "two great ideas" was the development of computer science. What Bronowski did not know was the circumstance that had inspired this. For von Neumann's insights at Los Alamos inspired the breakthrough to a viable thermonuclear bomb, and it was to complete that task that he developed and applied the world's first electronic digital computers, the ENIAC and MANIAC—the Electronic, or respectively Mathematical Analyser, Numerical Integrator And Calculator.

Recall his first great secret success had been to solve the mathematics of implosion but for which the plutonium bomb might never have worked. The Trinity test and the devastating explosion over Nagasaki showed that von Neumann's analysis had been correct, but for the "thinking animal" this success inspired new questions and ideas. What if the centre of the plutonium shell contained a mix of deuterium and tritium?

The fission of plutonium releases neutrons in less than a shake. It takes several shakes for the whole shell to fission, during which these "early" neutrons heat the remaining plutonium and raise the temperature of the DT (deuterium and tritium) mixture within. The whole assembly ionises. Atom for atom plutonium has more ions than the DT mixture, which results in the plutonium squeezing the DT fuel and compressing it. A highly compressed DT gas is easier to ignite than a quiescent mix. This trick due to von Neumann is known as *ionisation compression*.

Once ignited, the deuterium and tritium fuse and produce fast neutrons which now stream outwards into the surrounding plutonium remnants. Recall how fission explosions disassemble the plutonium into subcritical pieces, causing the chain reaction to end, which limits the total power of a basic fission weapon. In von Neumann's design the neutrons spawned by the deuterium-tritium fusions blast into the plutonium shards, initiate a further generation of reactions, and significantly boost the power of the explosion.

This, the principle of boosted fission bombs, inspired Edward Teller to hope that by surrounding the DT mixture and the plutonium shell with a further sphere of fusion fuel, a full thermonuclear explosion would result in this outer shell. But as Fermi's calculations would reveal, the amount of tritium needed was impractical. That was the situation after Fermi's lectures in the autumn of 1945.

Fermi's analysis was a brilliant survey, but to have a full understanding of the complexity of these electrically charged gases of deuterium and tritium at high temperatures would require much more detailed mathematical investigations. These proved to be so difficult that theoretical work on the Super stalled. Von Neumann decided to solve the equations using a computer.

Computers were at that time primitive mechanical devices, with gears and many moving parts. In 1944 came the ENIAC, a machine using vacuum tubes. Von Neumann became excited by this, studied its engineering, and worked out how to use its logical system for processing information. This work, in 1945, published after the war ended and widely regarded as the most important ever on computing and computers is what Bronowski referred to as one of von Neumann's great ideas. Much of the modern jargon—operating systems, hardware, software—derive from his insights.

What was not publicly advertised was that von Neumann's computers were designed to solve the problem of energy flow in the hellish DT fuel of the hydrogen bomb. By 1946 the results of his computations confirmed Fermi's broad conclusions about the Classical Super's nonviability, much to Teller's chagrin. This led to a secret three-day conference at Los Alamos in April 1946, during which Fuchs had an idea how to build on von Neumann's invention of the boosted fission bomb and ignite a genuine thermonuclear explosion.

First, replace the plutonium by uranium. The high explosives in the plutonium implosion absorbed many of the neutrons, while a uranium gun mechanism with no need for extraneous gunpowder would not. So, one subcritical uranium hollow sphere filled with DT fuel, à la von Neumann, is hit by a subcritical lump of fast-moving uranium blasted from a gun. The impact makes a supercritical mass that explodes. So far, this is like von Neumann's boosted fission bomb, but with uranium-235 instead of plutonium. Another key difference is that, thanks to the momentum imparted by the gun, the exploding material is in motion. Fuchs's brainwave was to have another capsule of deuterium and tritium, enclosed in a tamper of solid beryllium oxide, remote downstream, which the exploding bomb is now hurtling towards at the speed of sound. The clever insight is that X-rays from the explosion are travelling even faster—the speed of light—and hit the waiting capsule momentarily earlier.

The impact of the radiation exerts a pressure on the capsule, ionises it, and causes it to implode. This *radiation compression* speeds up the next stage of the explosion when the real compression occurs—the one-two punch of the radiation compression followed by the ram's brute force. This combination of radiation at light speed followed by the impact of fission debris is the key ingredient to ionisation compression, which would become integral to subsequent designs of thermonuclear weapons.

But not yet. The computations implied it would be ideal kindling to start a thermonuclear fire, but the problem of maintaining the explosion remained. That this was indeed a much more efficient trigger for igniting a fire of deuterium was proved in the Operation Greenhouse George test of 1951 at the Enewetak Atoll in the South Pacific. The fission reaction was separated from the capsule containing deuterium and tritium, and a long pipe channelled radiation from the fission explosion to the capsule, causing it to ignite in the world's first thermonuclear blaze. Even so the DT capsule was unable to maintain burning in the Fuchs–von Neumann device. This was the first genuine ignition of a fusion explosion, but its effect overall was described as like using a blowtorch to light a match.

20
The Ulam–Teller Invention

When mathematician Stanislaw Ulam of the University of Wisconsin received the summons to Los Alamos in 1943, all he was told was that it was secret, interesting, and important. Apart from that he had nothing other than a rail ticket to a station near Santa Fe in New Mexico. He found an atlas of New Mexico in the university library and noticed that the list of recent borrowers were physicists who had disappeared from his department. Upon arrival at his destination, only one other person alighted from the train. Ulam recognised his fellow traveller to be John von Neumann who, to his surprise, was greeted by their contact as "Mr Norman". Secret indeed.

During the war, Ulam invented a way of making approximate analysis of otherwise insoluble problems, today known as the Monte Carlo method.[1] This was key to much of the analysis done by von Neumann,

Peierls, and Fuchs, and many others, during the atomic bomb design. After the war, Ulam stayed on at Los Alamos and turned his attention to the hydrogen bomb.

After the April 1946 conference, Edward Teller made two breakthroughs, neither of which at the time solved the problem of how to make the Super work but contributed eventually to the final successful design. In August 1946 he invented the Alarm Clock, which he hoped would reawaken interest in the Super. This was an extension of von Neumann's earlier idea of using fusion fuel to boost the power of a fission weapon; Teller proposed an onion-like structure consisting of alternating spherical layers of fissionable materials—such as uranium or plutonium—and thermonuclear fuel, with the goal of boosting a fusion bomb.

The outer layer consisted of uranium-238. It will fission when hit by fast neutrons, and these are produced by DT fusion taking place in the next inner shell of fuel. The idea was to compress the outer shell of uranium-238 by chemical explosives, then the neutrons from the resulting fission reactions would ignite fusion in the DT mixture and send a shock wave inwards to hit the next layer of this nuclear onion—another shell of uranium. Within this inner shell is the fusion fuel that Teller hoped would be compressed so much that it would then explode. The key point was that whereas the outer shell of uranium had been imploded by the means of *chemical* explosives, this inner shell would now receive the full force of a *nuclear* blast. His idea of using the force of both fission and fusion to beat down onto the central core at the heart of the device was good, but initiating the whole process at the remote surface with chemical explosives was quickly realised to be practically impossible.

In September 1947 Teller had his second idea on how to improve the design. As Fermi had already pointed out in his 1945 lectures, ignition is easier with tritium than deuterium, but tritium is hard to come by. Teller came up with a clever way to make tritium within the weapon itself. His novel approach was to use solid lithium deuteride as

The Ulam–Teller Invention

$$n + Li_6 \longrightarrow t + He_4$$

Figure 10. A neutron hitting lithium-6 produces tritium inside the bomb. Protons are shown as solid circles and neutrons as open circles.

fuel, where the key is to use the isotope lithium-6 (consisting of three protons and three neutrons) and the fact that when a neutron hits Li_6 it produces tritium (Figure 10). So, when neutrons from fission hit the lithium deuteride, they breed the fusion fuel right where it is needed.

For several months Ulam worked with Fermi making approximate calculations of the complicated mathematics to see whether Teller's ideas were likely to be enough to make a thermonuclear explosion. The nuclear physics was relatively straightforward: the fusion reaction between deuterium and tritium (Figure 9b in Chapter 19) releasing energy and a neutron is very simple; the chain reaction caused by that neutron impacting uranium or plutonium is more complicated but was well understood. The problem facing Ulam and Fermi is akin to knowing the flight of one starling when what is needed is to understand the murmuration of a flock of thousands. In the Super, the starlings correspond to individual atoms of deuterium, tritium, lithium, or uranium, the murmuration analogous to the shock waves of greater or lesser density within the fusion fuel.

Starlings are free to roam, forming artistic patterns, the edges of the flock rising and falling like waves. In the fusion weapon, making the shape spherical or cylindrical gives some control over the mathematical analysis. A cylinder, seen end on, is a circle repeated over and over; the sphere is a point growing the same in all three dimensions. The nuclear murmuration is constrained by symmetry. Even so, the mathematical analysis was slow and painstaking work, which von Neumann realised would be an ideal test for the new electronic computer. The result of his computations on ENIAC confirmed that the Classical Super wouldn't work. Alarm Clock and lithium deuteride notwithstanding, Teller's device would still fizzle, not burn.

Recall, although Fuchs and von Neumann's invention had been proven to ignite the DT mixture, the fire wouldn't spread. Basically, you can't get the cylinder of deuterium to burn because energy escapes faster than it is being replenished by fusion. Teller, mistakenly as it would turn out, even had a theorem that compression would not help. He concluded that while compression speeds up the burning, it also accelerates the cooling at the same rate. Everyone was stuck.

It was at this point that Ulam made his critical contributions. The equations describing the heat transfer within the fusion material revealed that if the compression was much stronger—maybe a factor of a hundred rather than ten—the material would become opaque. The effect would be that heating would occur within it, but radiation at the heart would be trapped and an equilibrium could result. In short, Teller's "theorem" was wrong: Ulam's computations implied that supercompression could enable the fusion fuel to burn and a thermonuclear explosion could be achieved. So much for the mathematics, but how can you supercompress the material in practice?

In December 1950, Ulam saw a way to increase the implosive force and thereby the compression. The shock wave of neutrons produced by an atomic bomb—the primary trigger—would set off a second bomb. This was initially an idea to enhance the power of atomic bombs which, politically, was the primary goal at that time. Then, in January 1951, he saw a way to apply it to the Super. The Classical Super was a system in which uncompressed deuterium would be heated; his idea of using staged fission explosions between spatially separated bombs seemed a way to achieve the goal of supercompressing the deuterium.

Initially Ulam's vision was based upon the shock wave of neutrons emitted by the first bomb hitting the second. Teller—possibly stimulated by Fuchs and von Neumann's insight of radiation implosion, which had just been proved to work at the Operation Greenhouse George test in October 1950—then said, radiation is the way. His point was that radiation from the first bomb will precede the neutrons and prime the target.

The Ulam–Teller hydrogen bomb would involve three separate explosions, so fast that they appear as one. The primary is a fission bomb which ignites the secondary fusion bomb. This combination is the trigger and in effect is a boosted von Neumann bomb. The first blast ignites fission in the casing of the spatially separated fusion bomb. The fusion bomb mathematically approximates to a cylinder, built like a series of coaxial cables. Viewed end on, the centre of the circular cross section is a "spark plug" of uranium-235 or plutonium. This is surrounded by a layer of lithium deuteride fusion fuel, which in turn is encased in a tamper of uranium-238 whose outer surface is then coated with lighter elements. The goal is that the blast from the trigger makes the lithium deuteride implode *before* it ignites. The need for this sequence is that heat makes things expand, and lithium deuteride is no exception. Expansion would release the compression, with the unwanted effect of breaking up everything before the fusion burning stabilises. The timescale for fission to occur is about one shake, which set the challenge of how to compress lithium deuteride in less than that. This is where some clever physics and geometry came into play.

The energy transmitted from the primary to the secondary bomb is mainly in X-rays. These travel 3 metres in one shake, which sets the scale for the size of the separation between the primary and secondary. This lead time, before the blast of the neutrons arrives, will be crucial.

The outer surface of the casing was made of light elements. When hit by the X-rays, this surface evaporates and is blown off with great force. Like a rocket engine—where fuel ejected in one direction thrusts the rocket in the opposite direction (Newton's third law of motion that action and reaction are equal and opposite)—this outward blast thrusts the inner part of the casing, made of uranium-238, inwards. This compresses its lithium deuteride contents. Meanwhile neutrons from the uranium-238 fission pour through the fuel. The key feature is that compression is slow enough that these neutrons arrive at the spark plug before much compression has occurred. If this timing is right, fission induced in the spark plug ignites the surrounding lithium deuteride from the *inside*. Meanwhile the lithium deuteride continues to be compressed on the *outside*. This supercompression is enough to make the lithium deuteride opaque and maintain the burning throughout the fuel.

For all this to work effectively, the radiation from the trigger must illuminate the casing uniformly. This is where some geometrical tricks—and due to their invention being classified even today, we can only say "probably"—come into play. Light waves emitted at the centre of a sphere will reflect from any point on the surface back to the centre. In an ellipse, there are two foci, where light emitted from one will reflect to the other. If the container is an ellipsoid—a three-dimensional analogue of an ellipse—then if the first bomb is put at one focus and the second at the other, reflected radiation from the first will illuminate the second from all directions. The outer casing is also heated by the radiation and emits a uniform glow of X-rays. The power from this *black body radiation* is directly proportional to the fourth power of the temperature, which at millions of degrees in the bomb is considerable. The details remain secret, but the general principles are known.

By 1951 von Neumann had improved the ENIAC computer's logical architecture. ENIAC was a wonderful calculator, but a one-trick pony: if you wanted a new calculation you had to physically rewire the machine, which involved reconnecting thousands of cables by hand. This meant that a single programming change could take days to implement. So was born MANIAC, which could store programs. Von Neumann came up with five basic systems which are what underpin computers even today: input and output mechanisms, a unit for memory, an arithmetic unit, and a central processor. MANIAC's first job was to perform the engineering calculations for building the hydrogen bomb. It ran for two months, twenty-four hours a day, and verified that—in theory—Ulam and Teller's invention should work.

With ironic timing it was on Halloween in 1952 that the United States made the first explosion of this deadliest of weapons, confirming that the Ulam–Teller device works in practice.[2] Code named "Ivy Mike", the 10-megaton blast took place on Elugelab, an island in the Enewetak Atoll of the South Pacific. The Nagasaki bomb was puny in comparison. For the monstrosity at Elugelab, Nagasaki was but the trigger, whose X-rays would ignite the main bomb. Three storeys high, filled with liquid deuterium and weighing 80 tonnes, the whole contraption looked more like an industrial site than a bomb.

The explosion instantly wiped Elugelab from the face of the earth and vaporised eighty million tonnes of coral. In their place was a crater a mile across, deep enough to hold fourteen buildings the size of the Pentagon, into which the waters of the Pacific Ocean poured. Scientists who were present at both the Trinity and Ivy Mike tests recalled that the flash was far beyond what they had witnessed with that atomic bomb. It was clear that something was horribly wrong as soldiers briefly saw bones as shadows, like a Röntgen X-ray image but of the whole body and taken at a range of 20 miles.

The mushroom cloud reached 80,000 feet in 2 minutes and continued to rise until it was four times higher than Mount Everest, stretching 60 miles across. The core was thirty times hotter than the heart of the sun, the fireball 3 miles wide. Whereas at Trinity the flash was over in a moment, the Ivy Mike hydrogen bomb kept pouring out heat ever more intensely as if the whole world was on fire. The sky shone like a red-hot furnace. For several minutes, many observers feared that the test was out of hand and that the whole atmosphere would ignite.

The Ulam–Teller design has been the basis of all hydrogen bombs since. The Ivy Mike explosion marked a real change in history, a moment when the world moved to a more dangerous path, for the hydrogen bomb is not just a more powerful weapon; it is a true destroyer of worlds. Herbert York, Director of the US weapons lab at

Livermore, reported: "Fission bombs, destructive as they might have been, were thought of [as] being limited in power. Now, it seemed, we had learned how . . . to build bombs whose power was boundless"—restricted only by the amount of fuel and length of the spark plug in the bomb.[3] The world didn't burn after the Ivy Mike explosion because the fuel was limited. On Edward Teller's blackboard at Los Alamos was a list of weapons with their abilities and properties displayed. "The last one on the list, the largest, the method of delivery was listed as 'Backyard'. Since that particular design would probably kill everyone on Earth, there was no use carting it anywhere."[4]

THE SPY IN THE SKY

It had taken the United States seven years from successful detonation of an atomic bomb to the 1952 explosion of the Ulam–Teller fusion bomb. The Soviet Union and the United Kingdom took only half the time. How was this done so efficiently?

To blame espionage is simplistic. The notorious Klaus Fuchs left Los Alamos in 1946 after his work on radiation implosion. Although he passed news of this to the Soviet Union, along with Teller's idea of using lithium deuteride after learning about it from him in 1948, by February 1950 Fuchs was in gaol.[5] He knew nothing of Ulam and Teller's breakthrough; that the Soviets had to (re)discover for themselves.

After the United States worked out how to make a hydrogen bomb and used one to vaporise an island in the Pacific Ocean, the fact couldn't remain a secret. What goes up must come down, which is what happened to the debris from Ivy Mike. Radioactive elements produced in the explosion shot into the stratosphere and were then spread around the globe to fall back to Earth. If there was any transfer of secrets to the Soviet Union, it was when the fallout over the USSR gave its nuclear scientists the clues to what had happened. There was enough information for knowledgeable nuclear physicists to reverse engineer the Ulam–Teller design.

The Ulam–Teller Invention

Among the fallout were elements never before found in nature. Today known as einsteinium and fermium, located at numbers 99 and 100 in the periodic table of elements, they are so far beyond uranium, neptunium, and plutonium as to be outside the Kuiper Belt.[6] Their creation in the exploding bomb showed that neutrons must have impacted heavy nuclei multiple times, undergoing a sequence of transitions to move elements far up the elemental ladder within a few shakes. For all this to have happened so quickly implied that at the heart of the hydrogen bomb large numbers of neutrons are created, their density raised markedly by compression before explosion. The variety of isotopes, and their abundance and distribution relative to one another, revealed the nuclear DNA and the structure of the weapon. That the Ulam–Teller device involved detonation of a fusion bomb by more than one fission bomb was clear, but the tricks which made the assembly work remained to be solved.

In 1954 another American test—Castle Bravo, at Bikini Atoll—included lithium deuteride in the fuel. This performed much better than anticipated and at 15 megatons was the most violent American bomb ever detonated. Much of its power came from the fission of its uranium-238 tamper. This created a huge amount of nuclear fallout, which unforeseen weather patterns blew across populated areas in the Marshall Islands over an area more than 100 miles long. The isotopes and elements detected in the fallout from Castle Bravo provided more forensic clues on how to build a thermonuclear weapon.

In the Soviet Union, Igor Kurchatov had successfully designed an atomic bomb. In September 1945 he received a memo from Yakov Frenkel, another Soviet physicist, suggesting that the high temperatures developed in an atomic bomb might "ignite nuclear synthesis reactions (such as production of helium from hydrogen) which are the energy sources of the stars and could add to the energy released in the explosions of the basic fissionable materials (uranium, bismuth, and lead)".[7] Although he erroneously thought bismuth and lead were fissionable materials, his basic idea was right and mirrored what Fermi had proposed to Teller in 1941. At about the same time, Kurchatov was told of Fuchs's news that the United States was working on a hydrogen bomb. These two events inspired the Soviets to investigate how to make such a weapon for themselves. Following Ivy Mike and Castle Bravo, this became an increasingly urgent quest.

Among the small team that Kurchatov organised was a brilliant young theoretical physicist: Andrei Dmitrievich Sakharov.

SAKHAROV'S SWEET IDEAS

In 1945 Andrei Sakharov was a twenty-four-year-old graduate student in Moscow at the theoretical physics institute of the Russian Academy of Sciences. Under the supervision of Igor Tamm, a leading Soviet physicist and future Nobel laureate, he researched nuclear transmutation. On the morning of 7 August 1945, on the way to a bakery, he saw displayed on a newspaper stand the announcement that an atomic bomb had been dropped on Hiroshima. His "knees buckled. Something new and terrible had entered our lives, and it had come from grand science—the one that [he] worshipped".[8]

Igor Kurchatov invited Sakharov to join the Soviet atomic bomb project in 1946, and again in 1947. Sakharov declined on both occasions as he wanted to continue research in fundamental science under Tamm's tutelage. He became a Doctor of Science in 1947.

Although at that time Kurchatov's focus was on building the Soviet *atomic* bomb, towards the end of 1945 he had become aware from Fuchs that the United States was interested in making a *thermonuclear* explosion in deuterium.[9] Kurchatov assigned Yulii Khariton, in collaboration with three colleagues from the Academy of Sciences chemical institute, to consider the possibility of using the energy of light nuclei to make a thermonuclear blast, and to report to a special committee chaired by Lavrentiy Beria, head of Soviet security. The effective lead theorist was Yakov Zeldovich.

While this work was going on, new intelligence reports arrived about the American H-bomb programme. In September 1947, Klaus

Fuchs—now back in England but continuing as a spy—met with Alexandr Feklisov, a Soviet intelligence officer, in London. Feklisov asked about the Super bomb. Fuchs said that the Americans were interested in tritium alongside deuterium, but he was unaware whether serious research and development were underway.

Feklisov met Fuchs again in March 1948. On this occasion Fuchs handed over material of paramount importance, including details of his work with von Neumann developing a trigger for the Classical Super. By April, a Russian translation of Fuchs's information was in the hands of the Soviet leadership, who decided emergency measures should be taken to intensify research into H-bombs.[10] Beria ordered Kurchatov and Khariton to analyse these materials and submit proposals on organising studies in light of the new intelligence.

Within forty-eight hours a special theory group was formed at the physics institute of the National Academy under the oversight of Tamm. The blanket of secrecy was so extreme, however, that Tamm had no access to the intelligence reports with their key data on tritium reactions that the Americans had identified would be key to a viable bomb. Beria denied Tamm sight of these materials because "this will enlarge the circle of persons familiar with them beyond necessity".[11] Armed with mere outline information about the American Classical Super, Tamm, Sakharov, and another student, Vitaly Ginzburg, began by considering a similar design which was nicknamed Truba—meaning "tube". Their task was to verify the calculations about deuterium detonation made by Zeldovich's group at the institute of chemical physics, and to improve them.

Sakharov analysed the Zeldovich calculations and, during a couple of months in the autumn of 1948, began thinking about an alternative solution to the problem. This, his "First Idea", determined a new direction for the entire project.[12]

Until then they had been considering the classic design where a uranium-235 explosion is the spark with liquid deuterium as the fuel. In October 1948, Sakharov had the brain wave: a heterogeneous structure with alternating layers of light elements (deuterium and tritium) and the common isotope of uranium, U238. This design was like Teller's Alarm Clock. Sakharov had come to this independently and called it Sloika—"layer cake" in English. The key feature of Sloika, as in Alarm Clock, was fission of heavy U238 inducing ionisation compression of the light fuel. Sakharov's colleagues dubbed it Sakharization—which not only gave him credit but created a pun linked to his technically sweet idea: *sakhar* is Russian for "sugar", its English cognate being *saccharin*.

Sakharov's breakthrough defined a new direction and rapidly led to his "Second Idea". This evolved from discussions in the group but is uniformly credited to Vitaly Ginzburg. On 2 December 1948, he proposed to "burn mixtures containing lithium-6 . . . to utilise the heat (energy) released in the reaction $n + Li_6 \rightarrow T + He_4 + $ [energy] along with U235, Pu etc."[13]

Ginzburg had stumbled on the same concept as Teller—using lithium-deuteride in the bomb—but his motivation was different. Ginzburg's idea to use lithium deuteride as a thermonuclear fuel was inspired by the amount of *energy* it produced. This contrasted Teller's motivation, which was to increase the *production of tritium* by the explosion. The reason Ginzburg didn't initially focus on production was because of Beria's obsessive internal secrecy: the tritium data sent by Fuchs showed tritium to be much easier to ignite than deuterium, but Ginzburg didn't know this key fact.

In March 1949 Ginzburg presented his report to Khariton, who assessed it. Khariton was party to the data on tritium, which showed the importance of using tritium in the fuel. This insight implied that breeding tritium in situ could be an even better reason for using Ginzburg's proposal. In effect, Khariton's insider knowledge had now brought the USSR to the point that Teller had reached in 1947. So, on 17 March Khariton requested of Beria that Tamm's group be allowed access to the intelligence data. Beria refused, but after the importance of the tritium data was explained to him, an agreement was reached:

Tamm could have tritium data but not be told where it had come from. The key facts were sent over on 27 April, which ironically was almost the same time that the Americans published the data, openly, in the *Physical Review*.[14]

Armed at last with essential information about tritium's advantages, Ginzburg updated his calculations, and the conclusion was clear: the Layer Cake armed with lithium deuteride was the way to go. As early as January 1949 Sakharov had suggested the use of an additional plutonium charge to initially compress the layer cake. This was a prototype of a two-stage thermonuclear charge. However, the difficulty was to find a straightforward way to implement the idea. A stumbling block was how to make a symmetric compression of the thermonuclear unit.

There was considerable doubt that a solution would be found, so research on both the Layer Cake and the Tube—the Russian name for the Classical Super—continued. The USSR lacked computers like MANIAC, which restricted their theoretical analysis of these complex systems of liquids and gases at very high temperatures and extreme pressures. Not until 1953 was Zeldovich's group finally able to prove that nuclear detonation in the Tube is impossible.

In June 1953, Sakharov, Tamm, and Zeldovich signed a report on how to develop a prototype weapon by using the first and second ideas: a layer cake of uranium-238 and fusion fuel, with lithium deuteride to generate tritium within the bomb. The test was planned to take place in the steppe of northeast Kazakhstan on 12 August. This would be the fourth in a series of thermonuclear bomb tests going back to 1949, leading to it being named in the West "Joe-4", after Joseph Stalin.

The bomb, small enough to be a deliverable weapon, was placed on a tower in the middle of the open grassland. Twenty miles away, Sakharov and young scientists from his and Zeldovich's groups lay on the ground facing the tower. Tamm, meanwhile, was invited to a bunker with the VIPs, who were "just as nervous as we were".[15] Stalin had died in March, which meant that failure would not necessarily lead to execution, though there remained plenty of labour camps in the Gulag ever ready to welcome new recruits.

The countdown began and was broadcast over loudspeakers: "ten minutes to go, five minutes, two minutes . . ." The watchers donned dark goggles. Sakharov recalled the tension: "60 seconds, thirty, ten, nine, eight, seven, six, five, four, three, two, one . . . we saw a flash and then a swiftly expanding white ball lit up the whole horizon". Sakharov took off his goggles and, though partially blinded by the glare, saw a "stupendous cloud trailing purple dusty streamers which turned grey, separated from the ground and rose upwards becoming a shimmering orange". The familiar shape of the mushroom cloud formed, its stem being much thicker than those of atomic bomb explosions. The cloud rapidly filled half of the sky, which "turned a sinister blue-black colour". The shock wave "blasted [his] ears and struck a sharp blow to [his] entire body".[16]

The Minister of Machine Building, Vyacheslav Malyshev, came out of the VIP bunker to say that the chairman of ministers, Georgy Malenkov, had phoned congratulations. Malyshev then joined Sakharov in a car to visit ground zero and see the results for themselves.

They drove to within about 50 metres of the site, then got out of the car and walked around "nonchalantly".[17] The land seemed to have turned to glass, forming a lunar landscape as far as the horizon in all directions. Foolishly, like Icarus, Sakharov and Malyshev had gone too close to an artificial sun, however. In early November Sakharov fell seriously ill with an undiagnosed blood disorder, which recurred at intervals though he managed to survive. Malyshev was not so fortunate, dying four years later of leukaemia at the age of fifty-four.

The 500-kiloton blast was dominantly caused by fission of uranium and only a fraction of the result of a genuine self-driven thermonuclear hydrogen bomb. To go beyond this would require a further breakthrough, a "Third Idea".[18]

News about America's powerful Castle Bravo test in March 1954 further stimulated developments in the Soviet Union. The explosion demonstrated that the Americans had found an answer, and fallout data gave some clues about their weapon's structure. After much thought and analysis of all the available information, Sakharov's team came to the answer by a process of elimination. They knew a single stage wouldn't work, which led them to focus on two-stage ignitions. Where previously this idea had focused on the mechanical shock wave of the primary fission debris as the intermediary, Sakharov now found a way to compress the secondary thermonuclear core by radiation from the primary nuclear charge. The key secrets of the Ulam–Teller mechanism were at last coming together in the Soviet Union.

Several intricate problems had to be resolved about the physical processes involved when radiation compresses matter, however. Sakharov found solutions to the mathematical partial differential equations, which confirmed that such an idea should work. This staged use of atomic explosion for imploding the thermonuclear fuel is what Sakharov refers to as the "Third Idea". Several of the team had brainstormed on blackboards, and from this collective the breakthrough emerged. Sakharov understood the idea's physical and mathematical aspects the deepest. This insight, together with the authority he had acquired through the years, enabled him to play the decisive role in its adoption and implementation.

Sakharov recognised that Zeldovich and others had undoubtedly made significant contributions and that they may have grasped the promise and the problems of the "Third Idea" as well as he had, but at that time they were all too busy to worry about who received credit. He later concluded: "Now it is too late to recall who said what during our discussions period and does it really matter that much?"[19]

The Soviet Union's first genuine two-stage hydrogen bomb—Joe-19—was tested successfully in November 1954. Sakharov was by now a full member of the Soviet Academy of Sciences, the winner of three Hero of Socialist Labour medals, as well as the Stalin Prize. In addition, he was given a luxurious dacha in the countryside near Moscow.

However, concerned about the implications of testing hydrogen bombs in the atmosphere, he began to lobby against them.

A fallout pattern forms like this. An explosion near ground level sucks up dust from the surroundings. This debris melts in the heat and in the process absorbs radioactive materials produced in the explosion of uranium and plutonium. The cloud blazes upwards, cools, mixes with the air, and is carried along with the winds. Heavier particles fall back to Earth first, lighter ones later, as the dust spreads out from ground zero. Sakharov realised this fallout could cause terrible damage to biological systems—not least humans—even to generations yet unborn. As experience from the tests grew, he became so concerned that he began to regard testing in the atmosphere as "a crime against humanity, no different from secretly pouring disease-producing microbes into a city's water supply".[20]

TOM, DICK, AND HARRY

The British H-bomb project was centred at Aldermaston, in the countryside about 40 miles west of London. Twenty miles northwest of Aldermaston, scientists at Harwell analysed the fallout from the Soviet tests.[21] This provided key information for the Aldermaston team, from which they were able also independently to discover the Ulam–Teller mechanism. The British exploded their first genuine hydrogen bomb on 28 April 1958, though the history is shrouded in misinformation and more secrecy than the American and Russian programmes.

Immediately after the end of the Second World War, the United States had cut its allies out of all nuclear work, both on atomic and hydrogen bombs, by the passing of the McMahon Act. In the United Kingdom, Prime Minister Winston Churchill decided that they had to go it alone. An atomic bomb was relatively straightforward as they could build on their experience in the Manhattan Project. This was done by 1952 under the direction of William Penney, a mathematician who had worked at Los Alamos but whose hands-on experience was not primarily nuclear physics. The presence of Klaus Fuchs as lead theorist, who was from 1947 to 1950 at Harwell, enabled the United Kingdom to develop a plutonium bomb like that tested at Trinity. Fuchs also alerted the Soviet Union of these atomic bomb ambitions.

Through the presence of its scientists at Los Alamos, the United Kingdom also knew of American interest in the hydrogen bomb, if only initially of the Super. When the United States made the Castle Bravo test in 1954, Churchill, once again Prime Minister, decided this new weapon was too dangerous to be left to America alone. To retain a seat at the table of world power, he decreed the United Kingdom must develop its own hydrogen bomb.

The sole British expert on the hydrogen bomb was Fuchs, whose activities by then had been traced and who was in gaol for espionage. Penney consulted him in prison and the basic elements to designing a hydrogen bomb, which Fuchs had already given to the USSR, were now given to the United Kingdom.[22] The British also had intelligence on the Soviet Union's 1953 tests, as well as the Castle Bravo American tests, thanks to analysis of the fallout made by the Harwell scientists. Armed also with Fuchs's idea of radiation implosion, the basic picture of a Ulam–Teller device was put together. Its detailed construction, especially the theory of detonation waves linking trigger and thermonuclear fuel, were a mystery. This is where the United Kingdom was faced with the same problem that had also tested Sakharov, Ulam, and Teller.

There were good scientists at Aldermaston but none of the stratospheric quality of Fermi, von Neumann, and Teller in the United States, or Sakharov, Tamm, and Zeldovich in the USSR. However, Lord Cherwell—still Churchill's science advisor—knew of John Ward, a remarkable young British theorist whose work on the foundations of the quantum theory of electromagnetic fields led Sakharov later to describe him as one of the titans of twentieth-century theoretical physics. Ward was at that time at Princeton, working in the same corridor as Einstein and von Neumann. Cherwell went over Penney's head in 1955 and recruited Ward to come to Aldermaston. Ward recalled his amazement at being "assigned the improbable job of uncovering the secret of the Ulam–Teller invention", the most carefully guarded of all US secrets and "an act of genius far beyond the talents of the personnel at Aldermaston".[23]

Ward was summoned to Penney's office with Keith Roberts, a theoretical physicist on the staff, briefed on what was known, and told to solve the problem with Roberts's assistance. Within six months he had done so; Roberts helped but Ward "worked almost entirely alone".[24] Penney, however, failed to appreciate the profound role of radiation implosion, and Ward, autistic and with a remarkable ability to alienate people, failed in the personal chemistry of bringing Penney onto his side. Penney dismissed the solution as unnecessarily complicated and overly expensive out of wartime.

Discouraged, Ward quit and returned to Princeton. In the short term this created security concerns for the United Kingdom as Ward was now back in America with full knowledge of the British hydrogen bomb programme. For Ward it was the start of a lifelong paranoia as he became increasingly convinced that the British security services were watching him. That the United Kingdom's programme didn't sink was thanks to Keith Roberts, who knew details of Ward's proposal, and the twenty-six-year-old Bryan Taylor, who had joined Aldermaston in August 1955.

Bryan Taylor developed the theory of a boosted fission weapon and analogues of the Alarm Clock or Layer Cake. The British initially designed a three-stage device. Nicknamed "Tom, Dick, and Harry", Tom, the primary, was a fission bomb to implode Dick, another fission device. Harry was the thermonuclear tertiary. In January 1956 they came up with a new design, involving only Dick (fission) and Harry (deuterium and tritium). At this point one of their colleagues, Ken Allen, realised that lithium-6 had been used. This led to the final British design: Tom, the fission trigger, and Dick, a set of concentric hollow spheres made of uranium interlaced with lithium deuteride. Roberts and Taylor completed John Ward's work on radiation implosion.[25]

The fathers of the hydrogen bomb in the United States and the Soviet Union are traditionally named as Teller and Sakharov, though many contributed along the way. In the United Kingdom, the four names known are those of Ward, Roberts, Taylor, and Allen; there is no clear claim of any of these to precedence, although in Ward's account he was the star.

The significance of Klaus Fuchs in this saga is beyond dispute because he played a key role in the United Kingdom's early work on its atomic bomb and was consulted while in gaol in 1954 by William Penney on his considerable knowledge about hydrogen bombs. Furthermore, he passed essential information to the Soviet Union, which both inspired their project and through his information on radiation implosion led to their independent discovery of the Ulam–Teller mechanism. And not least, in the United States itself, the Fuchs–von Neumann insight played a key role for Teller, if only subliminally, in his modification of Ulam's original staged trigger mechanism. While the names of the fathers of the hydrogen bomb in all these three separate endeavours may be the stuff of debate, Klaus Fuchs may be named unambiguously as the grandfather of them all.[26]

21
The MADness of Tsar Bomba

THE KING OF BOMBS

Andrei Sakharov called it the Big Bomb. The designers officially named it Big Ivan. In the West it has always been known as Tsar Bomba—the King of Bombs.

With a yield of 50 megatons, Tsar Bomba was the largest nuclear weapon ever constructed or detonated. More than two thousand times greater than the atomic bombs dropped over Japan, its single blast on 30 October 1961 exceeded the sum of the entire American Castle series of thermonuclear tests. Tsar Bomba has a unique place in history as the nuclear energy released in this single explosion amounts to 10 per cent of all known tests worldwide, before or since. Its 50 megatons are twice the thermal energy released in the most disastrous volcanic eruption in the United States in recorded history—Mount St. Helens in 1980. The volcanic explosion, however, was spread over several hours, whereas the blast of nuclear energy took less than a second: in terms of power, which is the rate that energy is released, Tsar Bomba was massively greater.

As if that were not madness enough, consider this: the original plan had been for the bomb to be 100 megatons, within a factor of two of Krakatoa's 200-megaton eruption. Concerned that this might have catastrophic effects, Sakharov directed the designers to cut the power in half. There was no scientific or military need for such a leviathan. The motivation was entirely political and President Nikita Khrushchev's need to appear tough.

Since 1958, the United States, the United Kingdom, and the Soviet Union had followed an informal moratorium on H-bomb tests. Khrushchev felt that the Warsaw Pact was vulnerable, however, due to the presence of American weapons in Europe, while he had none on the borders of the United States (which he tried to remedy with the Cuban Missile Crisis in 1962). He was also faced with a crisis in Berlin as East European citizens migrated freely to the more attractive West (which led to the construction of the Berlin Wall in August 1961). The United States meanwhile had a new president—John F. Kennedy—whom Khrushchev believed was weak. To have a bomb that would "hang over the heads of the capitalists, like a sword of Damocles",[1] Khrushchev ordered Sakharov to build a 100-megaton monster as part of a testing spectacular that would begin in September to coincide with the 22nd Congress of the Communist Party.

When Sakharov tried to explain that no further testing was necessary, and this news was passed up to Khrushchev, the president retorted, "Don't tell us how to behave. I understand politics. If we listened to people like Sakharov, I'd be a jellyfish not chairman of the Council of Ministers."[2] The whole purpose, for Khrushchev, was to make a spectacular demonstration of the Soviet Union's technical ability: there may be a shortage of plugs for wash basins, but the USSR can send Yuri Gagarin into space (as it had in 1961) and make big bangs. The mammoth weapon fitted with the Russian ethos: make it gigantic. The Kremlin was host to the world's largest bell—Tsar Kolokol—and cannon—Tsar Pushka—so Big Ivan became known in the West as Tsar Bomba.

From Khrushchev's order on 10 July to the deployment of Tsar Bomba in October was just sixteen weeks. The whole design process was rushed, the complex mathematics being circumvented by making shortcuts that degraded to little more than educated guesses. One theorist, Evsei Rabinovich, made some approximations which led him to conclude the device would fail. Sakharov responded by making different assumptions which led him to estimate—in modern slanguage, to guesstimate—that it could succeed.

The bomb, which was the size of a single-decker bus, was assembled in a special workshop straddling a railway line. After completion the carriage containing the weapon was disguised as a regular freight car and the workshop was dismantled. Construction took place while the concept was still on the drawing board, and there were serious doubts that it would work. The bottom line was that no one really had a clue. Sakharov was still worried by Rabinovich's concerns, so he introduced some changes in the design to reduce the chance of error. That night the engineers didn't go home but produced new blueprints. Sakharov reported to the ministry of the late-in-the-day change, which led the bureaucrats to panic as the test was due to coincide with the final session of the Congress, Khrushchev deliberately having chosen the timing to give the maximum psychological effect. One minister called in a rage: "Tomorrow I am due to fly out to the test site. What am I supposed to do? Call it off?" Sakharov calmed him down but remarked to colleagues that if the test failed "we will be sent to railway construction". On the other hand, their calculations implied that "if we succeed, this will open the possibility of creating a device of practically unlimited power"—Teller's "backyard" weapon.[3]

The calculations suggested that a 100-megaton bomb, built in three stages with thermonuclear fuel encased in uranium-238, would create so much fallout that northern Russia and even swathes of northwestern Europe would be contaminated. So, at Sakharov's urging, it was agreed to test the device at half strength by using lead instead of uranium as the fusion tamper. This eliminated the fast fission of the uranium by neutrons, which both reduced the yield by half and eliminated 97 per cent of the fallout. The result ironically was the "cleanest" weapon ever tested as 97 per cent of the energy came from fusion reactions. Had the original design been used, it would have instantly increased the world's total fallout since the invention of the atomic bomb by 25 per cent.

Tsar Bomba weighed 27 tonnes. Too big to fit inside the bomb bay of a regular plane, the device was slung beneath the fuselage of a specially adapted Tupolev Tu-95 long-range bomber. After taking off from an airstrip in the Kola Peninsula, in the far north of Russia, the Tupolev flew through cloudy skies over Novaya Zemlya, an archipelago in the Arctic Ocean. Accompanying the bomber was a second plane with a newsreel photographer to record the event. The chance of the planes' crews surviving the blast was rated as only 50 per cent.

At a height of 10,500 metres, the Tupolev released the bomb, its fall slowed by a parachute which itself weighed a tonne. Twenty-eight tonnes lighter, the plane leaped. The pilot banked it away from the forward momentum of the falling bomb and made their getaway at maximum throttle.

The device drifted down to about 4000 metres of altitude before it exploded. This was high enough that radioactive debris would not contact the ground and create significant local fallout, yet low enough that the planes would have time to escape far enough to avoid the storm. The blast wave bounced off the surface of the Earth and thrust the fireball of the bomb upwards. The mushroom cloud rose 65 kilometres, above the stratosphere and two thirds of the way to outer space. The mushroom's cap spread 100 kilometres from end to end. Almost all the fallout circled the northern latitudes for years.

The material damage was remarkable. There was complete destruction for 25 kilometres; 55 kilometres away in Severny Island all brick buildings were destroyed. This is analogous to both Gatwick and Heathrow airports being levelled by an explosion over central London. Hundreds of kilometres from ground zero, wooden houses were burned to cinders, stone ones losing their roofs, windows, and doors. A shock wave in the air was seen at Dikson Island 700 kilometres away; at 900 kilometres, windowpanes were broken. Atmospheric disturbance from the explosion lasted about four days, during which time it orbited Earth three times.

One cameraman wrote: "The clouds beneath the aircraft and in the distance were lit by a powerful flash. The sea of light spread under the hatch and even clouds began to glow and become transparent. After that moment, our aircraft emerged from between two cloud layers and down below in the gap a huge bright orange ball was emerging. . . . Slowly and silently, it crept upwards. Having broken through the thick layer of clouds it kept growing. It seemed to suck the whole earth into it. The spectacle was fantastic, unreal, supernatural."[4]

Scandinavian countries were horrified. The Soviet Union hadn't anticipated the enormity even though they had cut in half the original size. The flash was seen as far away as Norway, Greenland, and Alaska. About 1000 kilometres from ground zero "a powerful white flash burst over the horizon and after a long period of time [was] heard a remote, indistinct and heavy blow, as if the earth had been killed".[5]

Meanwhile Sakharov, far across the Soviet Union, sat by the phone on the day of the test waiting for news. He got confirmation that the plane was flying over the Barents Sea towards the drop zone. He couldn't focus on work, and colleagues hung around the corridor by his office, waiting for updates. At noon came a call: "There's been no communication with the test site or the plane for over an hour. Congratulations!"[6]

The lack of news was proof of success: the heat of the explosion had ionised the air, which like a huge solar storm hitting the ionosphere disrupted radio communications. Among the unanticipated consequences was the irony that contact with the command post on the Kola Peninsula and the scientists was immediately lost. The bigger the blast, the longer the blackout. Being incommunicado for an hour showed that the test explosion had occurred—and been huge.

A successful explosion was the goal, the health of the aircrew secondary, as another half an hour passed before the controllers received any reports on the test or the fate of the Tupolev bomber. Slowed by the parachute, Tsar Bomba had fallen for about three minutes before detonation, in which time the planes were nearly 50 kilometres away. The bomber was slower than the cameraman's plane and when the shock wave caught up, it was still only 115 kilometres away. The bomber dropped 1000 metres in the air, but the pilot was able to right it and land safely.

There is no military need for a 50-megaton device. Tsar Bomba was too heavy to be the warhead of an intercontinental ballistic missile, and for a conventional bomber to carry it to an enemy target would have required such a huge fuel load that the plane would probably have been unable even to take off, let alone fly. Had the bomber managed to reach its target with a 100-megaton device, the blast would have been so powerful that it would have been a one-way mission. The monstrosity was a machine of genocide, a city destroyer, an impossible and pointless weapon.

In 1940, Otto Frisch and Rudolf Peierls had discovered a means to extract nuclear energy explosively with power equivalent to that of "a thousand tons of dynamite".[7] A motivation was that an atomic bomb would be the only defence against an enemy who themselves had one. The philosophy of defence by balance of terror was born. Today the atomic bomb is no more than a thermonuclear detonator. The actual bombs that exploded over Japan in 1945 were equivalent to about 20 kilotons. In just twenty years Tsar Bomba, at 50 megatons, made the concept of Mutually Assured Destruction truly mad.[8]

NUCLEAR PARADISE LOST

The technical problems and industrial effort involved in making the atomic bomb were so huge that the advent of nuclear weapons might have been much delayed but for the needs of war. Had it not been for the unfortunate coincidence of fission's discovery and fear of an imminent collapse of society to fascism, nuclear power rather than nuclear weapons would have led the way. Postwar, Frederick Soddy's vision of nuclear energy "making the desert bloom" and the early postwar hopes of electricity "too cheap to meter" turned sour as fears of catastrophe spoiled public confidence in the promised utopian nuclear age.

Fission and fusion are respectively the keys to atomic and thermonuclear bombs. As atomic explosions proved to be relatively simple compared to thermonuclear, so has "conventional" nuclear power (fission) been with us for about eighty years, whereas fusion power is seemingly always forty years in the future.

The Manhattan Project included a team of primarily British and French scientists in Canada whose goal was to develop the science of nuclear reactors by building on Frédéric Joliot-Curie, Hans Halban, and Lew Kowarski's plans with the precious heavy water. After the war, American strategy was to develop hegemony in nuclear weapons. The so-called special relationship with the United Kingdom became less special following the US McMahon Act, which in 1946 ended cooperation on nuclear technology with the United States' former allies. The British and the French made their primary strategy the development of nuclear reactors to generate controlled nuclear energy for the benefits of society.

In the United Kingdom a century ago, what is today southern Oxfordshire sported apple orchards, green fields, and gallops for racehorses. Among its scattered settlements was Harwell, a place of half-timbered thatched cottages, with the downs crossing the landscape a few miles away to the south. The idyll was gone in 1946 when the site was taken over by an organisation referred to by locals as The Atomic—widely known simply as Harwell. This new organisation was in fact Britain's secret Atomic Energy Research Establishment, its goal to make the United Kingdom the world leader in the development of atomic energy. Within a decade it had done so. In addition to their analysis of nuclear fallout, on behalf of the UK secret bomb project, Harwell scientists and engineers had earlier designed and built the first experimental nuclear reactor, which "went critical" on 15 August 1947 and led to the world's first full-scale commercial nuclear power station, Calder Hall, opened by Queen Elizabeth II in 1956. Britain became the global leader in civil nuclear power.

This focus on nuclear reactors is long gone. Somewhere, the promise of Calder Hall and of UK leadership in nuclear energy fizzled out. France now dominates European nuclear energy production, leaving Britain far behind. It's notable perhaps that from 1978, just as France was launching the largest nuclear construction programme in European history, North Sea oil and gas were coming online. UK government policy seems to have been driven by a belief that gas would make the nation self-sufficient and that the supply would last forever. British nuclear energy now relies heavily on EDF (Électricité de France), a company that is 85 per cent owned by the French government. France, meanwhile, currently operates over fifty nuclear reactors, of which one site, Cattenom, just north of Metz, produces almost as much power as Britain's entire nuclear industry.

A tight secret, however, was that in the 1950s Britain also had its own independent atomic bomb project, and that atomic bombs made of plutonium rather than uranium were its main area of focus. So, while Harwell was developing the know-how to build nuclear reactors designed primarily to liberate atomic energy for peaceful uses, knowledge about the production of plutonium, its chemistry, metallurgy, and how to handle this nasty stuff was also accumulating.

The Faustian bargain of plutonium led also to a reactor at Windscale, today known as Sellafield, which produced fuel for the United Kingdom's independent atomic weapons programme. The Windscale plutonium reactor in Cumbria was a few hundred yards away from Calder Hall, but their different roles were a closely guarded secret. Only later did public suspicion grow that Calder Hall's commercial power programme was a cover for the primary purpose of developing weapons-grade plutonium.

On 8 October 1957, a technician at the plutonium reactor pulled a switch too soon, which ignited a fire. By the time it was extinguished, three days later, winds had spread radioactive fallout across the United Kingdom and the rest of Europe. The demonstration of what a nuclear accident could do, combined with the Macmillan government's attempts to censor the news and play down events, plus further news of radioactive leaks, increased scepticism.

The scientific progress in nuclear technology that was made in Britain during the 1950s has shaped the modern world. There are currently more than 400 operable power reactors across the globe with 160 under construction or planned. But for history, the number could have been higher. The Three Mile Island disaster of 1979 and the Chernobyl meltdown of 1986 have made the public and politicians instinctively wary of nuclear energy. More recently, the 2011 Fukushima disaster had profound effects on European energy policy, particularly in Germany, which turned sharply against nuclear power in response, opting instead to increase its reliance on Russian gas. And yet, the total numbers of people killed directly or indirectly by nuclear energy (such as from radiation-induced cancer) are miniscule compared with those who have suffered death or serious illness because of conventional power production. Accidents in mining coal, transporting materials, and processing fuel, and chronic bronchial conditions and worse from the stations' noxious emissions, seem to have been noticed less by the public than those linked to nuclear power.

Conventional nuclear reactors tend to be large and expensive, take a long time to build, and be notoriously plagued by cost overruns. The product might be too cheap to meter, but the cost of the infrastructure is immense. By the end of the twentieth century neither the outgoing Conservative government of John Major nor the incoming Labour government of Tony Blair saw an economic case for building new nuclear power stations. The French, by contrast, saw nuclear energy as the means for self-sufficiency.

Another of Harwell's nuclear children is also attracting much public attention. The peaceful production of energy from fusion, the power of the sun, was first explored at Harwell by researchers in the 1940s and in the USSR by Sakharov and Tamm in 1950. Today fusion is perceived as a potentially limitless, clean source of energy, but though it might be a relatively straightforward process in terms of physics, commercial fusion energy has proved very difficult to realise. A further complication is that fusion uses heavy isotopes of hydrogen—deuterium and tritium. As we have seen, tritium does not occur naturally and first must be made, like plutonium, in a nuclear reactor. So not that clean, then. And it is not widely realised that the origins of interest in "peaceful" fusion energy lie in the quest for nuclear weapons—in the United Kingdom for its own atomic bomb and in the USSR for the hydrogen bomb.

The energy of the neutrons produced in the fusion of deuterium and tritium makes them especially effective for converting uranium to plutonium. In 1946 this was anticipated to be both more efficient and easier to achieve than to construct a huge nuclear power plant to produce this essential fuel for the British atomic bomb. It was soon realised that fusion, while relatively simple physically, is a huge engineering challenge. Windscale was built to produce weapons-grade plutonium and interest in fusion gave way to the modern quest for the commercial production of energy.

In the Soviet Union, meanwhile, Sakharov had been trying to make improved designs for the Classical Super bomb. To do so he investigated the possibility of magnetic fields confining the ions of the thermonuclear fuel. Fermi too discussed a similar idea in 1945 in his Los Alamos lectures. The use of magnetic fields didn't solve the problem of the bomb but gave Sakharov the idea of the tokamak—a doughnut-shaped device, or torus, where magnetic fields control the behaviour of deuterium and tritium ions. By heating the fuel to some 200 million degrees, the deuterium and tritium would fuse and liberate power.

This is the basis of much fusion energy research today. Five miles from Harwell, an associated campus at Culham was for decades the centre for Europe's collaboration on building a realistic fusion power plant. JET—the Joint European Torus—operated there from 1983 to 2024.

Fusion has great potential—in theory. Experiments suggest that Europe is moving towards the possibility of a reactor capable of creating realistic amounts of usable fusion energy, and the culmination of this work will be the construction of the International Tokamak Experimental Reactor, the first fusion test reactor. This is hoped to begin operation in 2025. The location for this new reactor will be at Cadarache, in France.

NUCLEAR SHADOWS

Sakharov was so concerned by Khruschev's desire to go beyond Tsar Bomba, with potential devastating effects for the planet, that he argued powerfully for an international treaty banning all atmospheric testing. His stature and integrity raised awareness worldwide of the danger of the nuclear arms race. His efforts led to the signing of the 1963 nuclear test ban treaty. In the Soviet Union, however, he became a political outcast. This Hero of Socialist Labour and winner of the Stalin Prize for his development of the hydrogen bomb became ostracised in his homeland as he now spoke out against it.

Edward Teller's reaction to big bombs was very different. He seemed to have a mania for pursuing a route to the "backyard" weapon. Tsar Bomba was the biggest weapon to be exploded, but it was far from the largest that was seriously considered by either the Soviets, as we have seen, or the Americans. In 1954 Teller met with the general advisory committee of the United States Atomic Energy Commission and put forward ideas for weapons in the range of gigatons—thousands of megatons.

The minutes of the meeting show the committee's astonishment where Isidor Rabi said that it was absurd. In his opinion Teller was just trying an advertising stunt. Rabi remarked that in his estimation such an explosion could "set all of New England on fire. Or most of California. Or all of the UK *and* Ireland. Or all of France. Or all of Germany. Or both North *and* South Korea". Such a device, Rabi said, "might become a weapon of genocide".[9]

Was Teller serious? In Rabi's opinion, yes: "Until the next enthusiasm took over."

It appears that Teller's proposal was taken seriously by weapons developers in the United States as the titles, if not the classified content, of papers on projects Gnomon and Sundial reveal. Gnomon was a design for a 1 gigaton weapon, which would be the trigger to ignite Sundial—at 10 gigatons.

The Sundial bomb would have weighed over 1500 tonnes. The idea seems to have been to make the device the size of a submarine as the title of one document from 1956 reads: "Possibility of creating a tsunami from detonation of high yield weapons on the surface of deep water." This 10-gigaton monster is not yet Teller's "backyard" weapon, but that such a concept was ever considered is terrifying.

In 1896, the first pointer to nuclear energy had been so insignificant that it was almost missed. Within seven decades this had led to the explosive power of Tsar Bomba, the greatest ever recorded other than the meteorite impact sixty-five million years ago that wreaked global change and killed the dinosaurs.[10] The dinosaurs had ruled for 150 million years. Within just one per cent of that time, humans have produced nuclear arsenals capable of replicating such levels of destruction. The explosion of a gigaton weapon would signal the end of history. Its mushroom cloud ascending towards outer space would be humanity's final vision.[11]

POSTSCRIPT
A Nobel Trinity—Hahn, Rotblat, and Sakharov

OTTO HAHN

In 1938, Enrico Fermi had won the Nobel Prize for the wrong discovery. Credited with discovering transuranic elements, he had unknowingly produced fission of uranium.

Had the Swedish Academy taken more note of Ida Noddack's insight that Fermi had not eliminated the possibility of fission, they might have cited the true discovery. Just a week after Fermi was awarded his prize, Otto Hahn and Fritz Strassmann confirmed nuclear fission, and Lise Meitner and Otto Frisch realised its significance. Some seven years later, in August 1945, the awful consequences of fission were demonstrated in the explosion of atomic bombs over Japan.

By December 1945, around the fiftieth anniversary of the discoveries that had started this whole saga, Fermi had realised how even more devastating consequences could result from the control of nuclear energy with the development of thermonuclear weapons—hydrogen bombs—that could literally be destroyers of worlds. It is ironic that a week later the Swedish Academy awarded the still-vacant 1944 Nobel Prize for chemistry to Hahn for "his discovery of the fission of heavy nuclei".[1] There was no recognition for Meitner's decades-long work with Hahn that prepared the way for this great discovery, nor that it was she and Frisch, her nephew, who first understood its implications.

Meitner's gender has been suggested as a reason for her omission, but that would not explain why Frisch wasn't recognised. The significant numbers of Jewish scientists that won Nobel Prizes gives no support for a thesis that antisemitism played a decisive role in the selection process. However, antisemitic forces had clearly created the circumstances that would prevent Meitner being physically present when Hahn completed their experiment. She had fled Germany to escape the Nazis and was thereby forced to consult and direct Hahn from afar. Had she been able to remain in Berlin, she would have been inseparable from Hahn when awarding full credit for the joint discovery, though the key conversation with Frisch that revealed its fuller significance might then never have taken place. In any event, the history in my judgment shows that she should have shared the prize with Hahn, or with Frisch, or been one third of a trinity with them both. Instead, she was overlooked twice.

A final intrigue is the 1944 date for Hahn's prize. During the selection process in 1944, the Nobel Committee for Chemistry had decided that none of the year's nominations—Hahn's included—met the criteria as outlined in the will of Alfred Nobel, and no prize was awarded. The following year, 1945, the Committee awarded the prize to a Finnish scientist called Artturi Virtanen, for work in agricultural chemistry, but they also decided that Hahn deserved a prize after all. The

Nobel website enigmatically explains: "Otto Hahn therefore received his Nobel Prize for 1944 one year later, in 1945".[2] The atomic explosions were the singular events that had happened in the interim and must have convinced the Nobel Committee that the phenomenon had been experimentally demonstrated.

When the award was announced, however, Hahn was one of ten captured German atomic scientists being held at Farm Hall in England. The primary goal of the Allies was to determine how close the Nazis had come to making an atomic bomb. Hahn's whereabouts at that time were secret, and the Swedish Academy couldn't contact him. Instead, he learned of the prize when he read the news in the *Daily Telegraph*. The scientists were not released until 1946, so Hahn finally received his award only in November 1946.

Conversations among the interned scientists had been monitored by the British Intelligence Services, which revealed that Nazi Germany had achieved little beyond ideas for nuclear power or an undeliverable huge uranium bomb. They had never realised the opportunity of uranium-235 enrichment nor of plutonium, the bugging revealing that the German scientists were confused at what Pluto had to do with an atomic bomb. After all the fear that had driven the Manhattan Project, only now was it finally confirmed that the Nazis had utterly failed to develop a weapon. When one of the internees claimed that they would have made good if they had wanted Germany to win, Hahn had replied: "I don't believe that, but I am thankful we didn't succeed."[3]

JOSEPH ROTBLAT

For all the agonising of the scientists who had seen the awful consequences of their work, Joseph Rotblat was the only one who left the Manhattan Project for moral reasons when it became clear that its justification no longer applied.

Rotblat regarded the only purpose of the work at Los Alamos to be to build an atomic bomb as a deterrent to a possible German weapon.

By 1944, however, he realised that the war in Europe would almost certainly be over before the atomic bomb project was completed. Consequently, his participation in its development had become logically pointless. His unease increased following a dinner party at James Chadwick's house in March 1944. Leslie Groves was present and when Rotblat raised his concerns about the purpose of the Manhattan Project now that Germany was facing defeat, Groves responded tartly: "You realize, of course, that the whole purpose of this project is to subdue our main enemy, the Russians."[4]

In November 1944, Chadwick told Rotblat that intelligence information informed him that Germany was not working on an atomic bomb project. Rotblat immediately asked to quit the project and return to the United Kingdom. Groves and Los Alamos security immediately convinced themselves that Rotblat was a spy, but he was allowed to leave subject to not telling anyone his reasons.

He returned to the University of Liverpool as head of nuclear physics. However, wanting to use his skills for the benefit of mankind, in 1949 he became a professor of physics at Saint Bartholomew's Hospital medical centre in London University.

In 1957, as testing of nuclear weapons accelerated, Rotblat was a founder of the Pugwash Conference on Science and World Affairs. Pugwash had three main goals: education of the dangers of nuclear energy in war and peace, control of nuclear weapons, and responsibility of science to society. During the Cold War, Pugwash became a channel of communication between the communist Eastern bloc and the Western democracies. It played an important role behind the scenes in bringing about the nuclear test ban and nonproliferation treaties.

In 1995, a century after Wilhelm Röntgen's seminal discovery and fifty years after the nuclear age began, the Nobel Peace Prize was awarded to Rotblat and the Pugwash Conferences "for their efforts to diminish the part played by nuclear arms in international politics and, in the longer run, to eliminate such arms".[5]

ANDREI DMITRIEVICH SAKHAROV

Sakharov uniquely won two prizes linked to the hydrogen bomb: the Stalin Prize in 1953 for successfully creating the weapon, and the Nobel Peace Prize in 1975, in part for his work disowning it.

Tsar Bomba's 50-megaton blast in 1961 was the culmination of Sakharov's work on thermonuclear explosions. Khrushchev now wanted a demonstration of the full 100-megaton device. Sakharov was extremely worried. The replacement of uranium-238 by lead had reduced not just the bomb's power by half but critically eliminated most of its lethal radioactive fallout. A 100-megaton weapon could not only level urban areas in a zone 50 miles wide and cause third-degree burns in a region 120 miles across—in effect all southern England, from London to Bristol, and from Southampton to Birmingham—but also lethal fallout could extend into Warsaw Pact countries, while prevailing westerly winds would spread radioactive dust to Russia.

There was no practical means to deliver such a heavy weapon over an enemy target, which was just as well since there could be no moral justification for murdering tens of millions in the blink of an eye. Sakharov had a quandary, not dissimilar to that faced by Rotblat two decades before. In 1944, the atomic bomb would not be deliverable before Germany was defeated so there was no logical reason to pursue it nor, with the Nazis defeated, was there for Rotblat any moral reason. As a weapon of war, a 100-megaton bomb would be literally undeliverable to enemy territory; as for morality, the act of exploding such a monster in a test over the vastness of the Arctic would threaten humanity globally. Sakharov was concerned that the mushroom cloud might be repelled by the bomb's own blast wave, which could cause global fallout, spreading toxic dust around the planet with consequences both immediately and for centuries to come. Developing such a bomb had become for Sakharov both morally bankrupt and logically vacuous.

Concerned at the implications of his work for the future of humankind, Sakharov sought to raise awareness of the dangers of the nuclear arms race. He became a courageous activist for peace and disarmament, and for human rights. His efforts were partly successful with the signing of the 1963 nuclear test ban treaty. In the Soviet Union he was seen as a subversive dissident and was exiled to Gorky to limit his contact with foreigners. He founded a committee to defend human rights and the victims of political trials, and despite increasing pressure from the government he became one of the regime's most courageous critics, embodying the crusade against the denial of human fundamental rights. The 1975 Nobel Peace Prize citation reads: "For his struggle for human rights in the Soviet Union, for disarmament and cooperation between all nations".[6]

Sakharov was unable to be present at the award in Oslo which was accepted on his behalf by his wife Yelena Bonner. In a letter from Sakharov, which she read at the ceremony, he explained that he had been forbidden by the USSR to travel to Oslo "on the alleged grounds that I am acquainted with state and military secrets".[7] She explained that he was currently in Vilnius, capital of Lithuania, where a scientist colleague of his was on trial. She added that Sakharov was near the court building, not inside but standing out in the street in the cold, for the second day, awaiting the sentence against his closest friend.

In 1988, the European Parliament created the Sakharov Prize for Freedom of Thought. The prize that bears his name today goes far beyond borders to reward human rights activists and dissidents all over the world.

AFTERWORD

Wilhelm Röntgen claimed no patents for his discovery of X-rays as he wanted society to reap the benefits, and he donated his Nobel Prize money to the University of Wurzburg, for research. He was bankrupted by inflation after the First World War, and died of colorectal cancer in 1923, aged seventy-seven.

Henri Becquerel died of a heart attack in 1908, aged fifty-five. He was already suffering serious skin burns from handling radioactive materials, from which he had the idea of using radioactivity for radiotherapy.

Pierre Curie died in an accident in 1906, aged forty-six, but was already ill from the effects of radiation. Marie Curie died in 1935, aged sixty-six, after suffering from aplastic anaemia, caused by radiation damage in her bone marrow. She and Pierre were so contaminated by radiation that their tomb in the Pantheon is lined with lead.

Ernest Rutherford died in 1937, aged sixty-six, from a strangulated hernia. His ashes are interred in Westminster Abbey.

J. J. Thomson was president of the Royal Society from 1915 to 1920 and master of Trinity College Cambridge from 1918 to his death in 1940, aged eighty-three. His ashes are also interred in Westminster Abbey near those of Rutherford.

Frederick Soddy, after his return from Canada to the United Kingdom in 1903, became a professor at Glasgow, then Aberdeen, and finally Oxford University. He won the Nobel Prize for chemistry in 1921 for his work on isotopes. A polymath, he also had theories on economics and finance. He died in Brighton, aged seventy-nine, in 1956.

Hans Geiger worked on the German uranium programme during the Second World War. He endured the Battle of Berlin and Soviet occupation, dying in Potsdam in 1945, aged sixty-two.

Ernest Marsden moved to New Zealand in 1915 and became a bureaucrat in the Department of Education. During the Second World War he worked on radar, and postwar he studied the environmental effects of nuclear fallout. He died aged eighty-one in 1970.

Niels Bohr inspired the creation of CERN, the European Centre for Nuclear Research, to stem a brain-drain to the United States. He hosted its theoretical physics division in Copenhagen until it relocated to Geneva. He founded the Danish Atomic Energy Commission and died aged seventy-seven in 1962.

Otto Hahn also worked on the German uranium programme. After the war, he rebuilt German science and founded the Max Planck Society. He died in 1968, aged eighty-nine, after a fall.

Lise Meitner became a Swedish citizen in 1949 and settled in England in 1960 where her relatives lived. She was nominated for the Nobel Prize thirty times between 1937 and 1967. She died in her sleep after a stroke, three months after Hahn, aged eighty-nine.

Irène Joliot-Curie became a commissioner of the CEA (Commissariat à l'Energie Atomique, or the French Atomic Energy Commission) after the war and a pioneer of French nuclear power. She actively promoted the education of women. She died in 1956, aged fifty-eight, of acute leukaemia brought on by contact with radioactive materials.

Frédéric Joliot-Curie became the director of the CNRS (Centre National de la Recherche Scientifique, or the French National Centre for Scientific Research) and the Commissioner for Atomic Energy in France. His communist affiliations combined with political paranoia led him to be sacked from the CEA in 1950, but he retained a professorship at the Collège de France. He died in 1958, aged fifty-eight, from liver disease brought on by the effects of radiation.

James Chadwick became head of the British mission to the Manhattan Project during the Second World War. In 1948 he was appointed master of Gonville and Caius College, Cambridge. He died in 1974, aged eighty-two.

John Cockcroft founded the Harwell Atomic Energy Research Establishment in 1946 and was its first director. He became the first master of Churchill College, Cambridge, in 1959. He died in 1967, aged seventy.

Ernest Walton moved to Trinity College Dublin in 1934, spending his whole career there where he was renowned as a superb lecturer. He shared the 1951 Nobel Prize for physics with Cockcroft. He died in 1995, aged ninety-one.

Leo Szilard joined the University of Chicago in 1946 where he developed an interest in biology and founded the Salk Institute. He designed a treatment for cancer using cobalt-60, which he himself received to combat bladder cancer. He died of a heart attack in 1964 at the age of sixty-six.

Otto Frisch became head of nuclear physics at Harwell in the United Kingdom in 1946. He invented SWEEPNIK, a computerised system for measuring the tracks of atomic particles in bubble chambers. He became a professor of physics at Cambridge University and a fellow of Trinity College Cambridge, until his death, aged seventy-four, in 1979.

Rudolf Peierls, Frisch's colleague in "fathering" the atomic bomb, was the lead British theorist at Los Alamos, after the war returning to Birmingham University. In 1963 he became head of theoretical physics at Oxford University. Concerned with the nuclear weapons he had helped to unleash, he was president of the British Atomic Scientists Association and was actively involved with Joseph Rotblat in the Pugwash movement. He died in 1995, aged eighty-eight.

J. Robert Oppenheimer was chair of the US Atomic Energy Commission (USAEC) and argued against development of the H-bomb. His security clearance was infamously withdrawn in 1954. He was head of the Institute for Advanced Study in Princeton. A chain smoker, he died in 1967, aged sixty-two, from throat cancer.

Enrico Fermi was a member of the USAEC general advisory committee and supported Oppenheimer during the security hearings. He remained active in physics and made seminal contributions to the theory of high energy cosmic rays. In 1954, aged fifty-three, he died of stomach cancer, probably linked to his exposure to radiation in Rome.

Edward Teller was a tireless advocate of a strong nuclear weapons programme and became director of the Lawrence Livermore Laboratory. His testimony triggered Oppenheimer's downfall in 1954 and led to Teller being ostracised by many in the physics community. He developed schemes for using nuclear explosions in oil exploration and to neuter hurricanes, and he raised the ire of environmentalists with a proposal to use nuclear explosions to excavate a harbour in Alaska. He died aged ninety-eight in 2003.

Hans Bethe campaigned vigorously against nuclear weapons' testing. He published groundbreaking theoretical papers into his nineties on astrophysics, supernovae, solar neutrinos, and massive neutrinos. He died aged ninety-eight in 2005.

Stan Ulam proposed using nuclear power as a propulsion system for rockets. He developed electronic computing techniques to study problems statistically, known as the Monte Carlo method. He turned to biology and founded biomathematics. He died in 1984, aged seventy-five.

John von Neumann developed ICBMs—intercontinental ballistic missiles—and became lead defence scientist at the Pentagon. In 1957, aged fifty-three, he died of cancer that is widely believed to have been a result of exposure to radiation at Los Alamos.

Igor Kurchatov was involved in 1949 in a serious radiation accident which became a catastrophe at the Chelyabinsk-40 nuclear reactor.

Without proper safety gear he attempted to save the uranium and plutonium by stepping into the central hall of the damaged reactor which was full of radioactive gases. His health progressively declined after a stroke in 1954, and he died in 1960 at the age of fifty-seven.

Andrei Sakharov's activism against nuclear testing and the Soviet government led him to internal exile from 1980 to 1986. After being released by President Gorbachev, in 1988 he became a member of congress and leader of the opposition. He died in 1989, aged seventy-eight.

Vitaly Ginzburg made the first phenomenological model of superconductivity and won a Nobel Prize in 2003. After the collapse of the Soviet Union, he became a powerful supporter of human rights and warned against the return of Stalinism. He died in 2009, aged ninety-three.

Yakov Zeldovich became interested in fundamental theoretical problems and from 1952 concentrated on particle physics, cosmology, and astrophysics. In 1963 he returned to academia and worked on the thermodynamics of black holes and the structure of the large-scale universe. He died in 1987, aged seventy-three.

Yulii Khariton served as the scientific director of Arzamas, the Soviet Los Alamos, for forty-six years. Deemed by the Soviet authorities to be too important to risk flying, he had a personal railway carriage. He died in 1996, aged ninety-two.

Joseph Rotblat devoted the rest of his life to furthering the peaceful uses of nuclear physics and campaigned ceaselessly against nuclear weapons. As a professor at Saint Bartholomew's Hospital in London he studied the effects of radiation and of nuclear fallout on living organisms, and he became an advisor to the World Health Organisation. He was a leading critic of the nuclear arms race. He died in 2005, aged ninety-six. He never remarried.

BIBLIOGRAPHY

Albright, Joseph, and Kunstel, Marcia. *Bombshell: The Secret Story of America's Unknown Atomic Spy Conspiracy.* Times Books, 1997.

Amaldi, Edoardo. Interview with Thomas Kuhn. American Institute of Physics, 8 April 1963, www.aip.org/history-programs/niels-bohr-library/oral-histories/4484.

Amaldi, Edoardo. "Ettore Majorana, man and scientist". In *Strong and Weak Interactions: Present Problems,* edited by A. Zichichi, 10–17. Academic Press, 1966.

Amaldi, Edoardo. "Riccordo di Ettore Majorana". *Giornale di Fisica* vol. 9 (1968), 300.

Bethe, Hans. *The Road from Los Alamos.* American Institute of Physics, 1991.

Bickell, Lennard. *The Deadly Element: The Story of Uranium.* Macmillan, 1980.

Bohr, Niels. "Neutron capture and nuclear constitution". *Nature* vol. 137 (1936), 344–348.

Bohr, Niels. "Transmutation of atomic nuclei". *Science* vol. 86 (1937), 161–165.

Brown, Andrew. *The Neutron and the Bomb: A Biography of Sir James Chadwick.* Oxford University Press, 1997.

Campbell, John. *Rutherford: Scientist Supreme.* AAS Publications, 1999.

Cathcart, Brian. *The Fly in the Cathedral: How a Small Group of Cambridge Scientists Won the Race to Split the Atom.* Viking, 2004.

Close, Frank. *Neutrino.* Oxford University Press, 2010.

Bibliography

Close, Frank. *Half Life: The Divided Life of Bruno Pontecorvo, Physicist or Spy*. Basic Books, 2015.

Close, Frank. *Trinity: The Treachery and Pursuit of the Most Dangerous Spy in History*. Penguin Books, 2020.

Dombey, Norman. "John Clive Ward, 1 August 1924–6 May 2000". Biographical Memoirs of Fellows of the Royal Society, 17 February 2020, https://royalsocietypublishing.org/doi/10.1098/rsbm.2020.0023.

Emling, Shelley. *Marie Curie and Her Daughters: The Private Lives of Science's First Family*. Palgrave Macmillan, 2012.

Esposito, Salvatore. *Ettore Majorana: Unveiled Genius and Endless Mysteries*. Translated by Laura Gentile de Fraia. Springer, 2017.

Farmelo, Graham. *Churchill's Bomb: A Hidden History of Science, War and Politics*. Faber and Faber, 2013.

Fermi, Enrico. *Super Lectures nos.1–6*. Notes by D. R. Inglis of Enrico Fermi's classified lectures at Los Alamos, August to October 1945. LA-344. Los Alamos National Laboratory, via Freedom of Information request FOIA 09-00015-H.

Fermi, Laura. *Atoms in the Family: My Life with Enrico Fermi*. University of Chicago Press, 1954.

Frisch, Otto. *What Little I Remember*. Cambridge University Press, 1979.

Goncharov, German A. "American and Soviet H-bomb development programmes: Historical background". *Physics-Uspekhi* vol. 39 (1996), 1033–1044.

Goncharov, German A. "On the history of creation of the Soviet hydrogen bomb". *Physics-Uspekhi* vol. 40, no. 8 (1997), 859–867.

Goodman, Michael S. "The grandfather of the hydrogen bomb?: Anglo-American intelligence and Klaus Fuchs". *Historical Studies in the Physical and Biological Sciences* vol. 34, no. 1 (2003), 1–22.

Gowing, Margaret. *Britain and Atomic Energy, 1939–1945*. Palgrave Macmillan, 1964.

Guerra, Francesco, Leone, Matteo, and Robotti, Nadia. "The discovery of artificial radioactivity". Preprint, 2012. Available at https://iris.unito.it/retrieve/handle/2318/131329/14595/686404_Matteo_LEONE_The%20Discovery%20of%20Artificial%20Radioactivity.pdf.

Hahn, Otto. *New Atoms: Progress and Some Memories*. Elsevier, 1950.

Hirsch, Daniel, and Mathews, William G. "The H-bomb: Who really gave away the secret?" *Soviet Physics Uspekhi* vol. 34, no. 5 (1991), 437–443.

Bibliography

Holloway, David. *Stalin and the Bomb: The Soviet Union and Atomic Energy, 1939–1956*. Yale University Press, 1994.

Kevles, Daniel J. *The Physicists: A History of a Scientific Community in North America*. Harvard University Press, 1987.

Marnham, Patrick. *Snake Dance: Journeys Beneath a Nuclear Sky*. Chatto and Windus, 2013.

Mercurio, Jed. *Ascent*. Simon and Schuster, 2007.

Oliphant, Mark. *Rutherford: Recollections of the Cambridge Days*. Elsevier, 1972.

Peierls, Rudolf. *Bird of Passage: Recollections of a Physicist*. Princeton University Press, 1985.

Recami, Erasmo. "Ettore Majorana: The scientist and the man". *International Journal of Modern Physics D* vol. 23, no. 14 (2014), 1444009.

Rhodes, Richard. *The Making of the Atomic Bomb*. Penguin, 1988.

Rhodes, Richard. *Dark Sun: The Making of the Hydrogen Bomb*. Simon and Schuster, 2005.

Sakharov, Andrei. *Memoirs*. Hutchinson, 1990.

Schwartz, David N. *The Last Man Who Knew Everything: The Life and Times of Enrico Fermi, Father of the Nuclear Age*. Basic Books, 2017.

Segrè, Emilio. *A Mind Always in Motion: The autobiography of Emilio Segrè*. University of California Press, 1993.

Segrè, Gino, and Hoerlin, Bettina. *The Pope of Physics: Enrico Fermi and the Birth of the Atomic Age*. Henry Holt and Co., 2016.

Serber, Robert. *The Los Alamos Primer: The First Lectures on How to Build an Atomic Bomb*. University of California Press, 1992.

Sime, Ruth Lewin. *Lise Meitner: A Life in Physics*. University of California Press, 1996.

Snow, C. P. *The Physicists: A Generation That Changed the World*. Macmillan, 1981.

Sobel, Dava. *The Elements of Marie Curie: How the Glow of Radium Lit a Path for Women in Science*. 4th Estate, 2024.

Soddy, Frederick. *The Interpretation of Radium*. John Murray, 1909.

Solvay, Institut International de Physique. *Structure et Propriétés des Noyaux Atomiques: Rapports et Discussions du Septième Conseil de Physique tenu á Bruxelles du 22 au 29 Octobre 1933*. Gauthier-Villars, 1934.

Szasz, Ferenc. *The Day the Sun Rose Twice: The Story of the Trinity Site Explosion July 16, 1945*. University of New Mexico Press, 1984.

Szasz, Ferenc M. *British Scientists and the Manhattan Project: The Los Alamos Years*. Macmillan, 1992.

Teller, Edward (with Judith Shoolery). *Memoirs: A Twentieth-Century Journey in Science and Politics*. Perseus, 2001.

Teller, Edward (with Allen Brown). *The Legacy of Hiroshima*. Macmillan, 1962.

Ward, J. C. *Memoirs of a Theoretical Physicist*. Edited by F. J. Duarte. Optics Journal, 2004. Available at https://goldbart.gatech.edu/PostScript/Ward-memoirs/JCWard-memoirs.pdf.

Wells, H. G. *The World Set Free*. Macmillan, 1914.

Wilson, David. *Rutherford: Simple Genius*. Hodder and Stoughton, 1983.

NOTES

Prelude: Trinity 1945

1. The weather conditions have been widely documented: e.g., Atomic Heritage Foundation, "Trinity test—1945", 18 June 2014, ahf.nuclearmuseum.org/ahf/history/trinity-test-1945/.

2. Although the plutonium was only months old, its nuclear energy derived from the pre-existing uranium.

3. Philip Morrison, quoted in Rhodes, *The Making of the Atomic Bomb*, p. 673.

4. A notable exception is Christopher Nolan's *Oppenheimer* (Universal Pictures, 2023), which used the period of silence to build tension.

5. If symmetry implies one revolution per century, the development of "intelligent" computers may now be shaping the Fourth: "Industry 4.0". Society is rapidly having to weigh the threats and opportunities of this emerging new age while still conflicted with those of the Third. I am a physicist, not a sociologist, and here focus on how the nuclear age came to be, not primarily on societal attitudes towards its benefits and hazards.

Chapter 1: The Third Revolution

1. The actual values are the weighted averages of masses and abundances of an element's isotopes (see Chapter 3) and so deviate from whole numbers.

2. Mendeleev's original 1869 version of the periodic table and its subsequent development are described in: Deboleena Guharay, "A brief history of the periodic table", *ASBMB Today*, 7 February 2021, www.asbmb.org/asbmb-todayscience/020721/a-brief-history-of-the-periodic-table.

3. Today determined as approximately 232 and 238 AMU, respectively.

4. *The Reason Why—in Science*, edited by J. Scott, M.A., Sisley's Ltd, Makers of Beautiful Books (undated).

5. Now understood to be due to atoms having inner structure, which enables them to lose or gain electrons and thereby become positively or negatively charged; see "What Is Electricity?" on page 22.

6. Attributed to W. Crookes, date unknown.

7. Röntgen's paper "Über eine neue Art von Strahlen" (On a new kind of rays) was read at the Wurzburg Physico-Medical Society on 28 December 1895. An Austrian newspaper reported this on 5 January 1896; public awareness seems to have arisen from that.

8. J. J. Thomson, "Cathode rays", *Philosophical Magazine and Journal of Science* vol. 5 (October 1987), p. 312.

Chapter 2: From New Zealand to the World
1. As told to Mark Oliphant and reported in Campbell, p. 192, and Wilson, p. 62.
2. J. Chadwick, "The Right Hon. Lord Rutherford of Nelson, O.M., F.R.S." (obituary), *Nature* vol. 140 (1937), pp. 749–751.
3. Oliphant, p. 4.
4. See Campbell, Chapter 1, for commentary on Rutherford's luck and early life, and pp. 185–192 for the machinations behind his scholarship award.
5. Campbell, p. 192.
6. E. Rutherford, "Uranium radiation and the electrical conduction produced by it", *The London, Edinburgh, and Dublin Philosophical Magazine and Journal of Science* 47 (1899), p. 116.
7. All electrons are identical. Those emitted in beta radioactivity are indistinguishable from those found in the outer reaches of atoms.
8. Rutherford's work on alpha and beta rays was completed at the Cavendish in 1898 but was published in 1899 after he had moved to Canada.
9. Wilson, p. 130.
10. These experiments and their limitations are described in Wilson, p. 131.
11. Rutherford quoted in Wilson, p. 133.
12. Wilson, p. 133; Soddy, p. 26.
13. We now know that X-rays are emitted by electrons that occur deep inside large atoms, whereas gamma, alpha, and beta rays all originate in the atomic nucleus.
14. Wilson, p. 135.
15. Wilson, p. 137.
16. Wilson, p. 138.
17. Substances with very long half-lives maintain radioactivity at substantially a constant rate. In such cases a measurement of the activity—disintegration rate—is made for a given mass of the substance. The latter determines the number of atoms, and when combined with the disintegration rate it is possible to calculate how long it will take for half of the atoms to decay. Substances with short half-lives reveal measurable changes in intensity directly.
18. Soddy to McGill University Physical Society, February 1902, quoted in Wilson, p. 154.
19. In 1900 the German chemist Friedrich Dorn discovered radon in sealed vessels of radium that had been separated from pitchblende by the Curies. Neither Dorn nor the Curies knew how the radon had got into those vessels. This saga illustrates the confusion surrounding radioactivity and the elements at that time. Rutherford and Soddy's emanation from thorium became initially known as "thoron", the name "radon" being associated with the analogous emanation associated with radium, and back then these were thought to be two distinct and novel elements. Today we understand them to be two different forms of the same chemical element: radon-225 and radon-222, the numbers reflecting their atomic masses. These two versions of the same element are known as *isotopes*, a concept introduced by Soddy in 1913. Frederick Soddy, "Intra-atomic charge", *Nature* vol. 92 (1913), pp. 399–400.
20. J. Chadwick in the foreword to Oliphant, p. ix.

21. Wilson, p. 256.
22. Soddy, p. 224.

Chapter 3: Otto Hahn and Lise Meitner
1. Soddy, pp. 224, 232, and 244.
2. Soddy, p. 4.
3. The name isotope was suggested by Glaswegian novelist Margaret Todd during a dinner party at Soddy's in-laws in Sobel, p.139.
4. Hahn, p. 146.
5. Boltwood's letter to Rutherford, quoted in Hahn, p. 146.
6. Folklore attributed to Rutherford, quoted without source in: Nicholas Amendolare and Kimberly Uptmor, "Otto Hahn—Biography and nuclear fission discovery", Study.com, 21 November 2023, https://study.com/academy/lesson/otto-hahn-inventions-nuclear-fission.html.
7. Hahn, p. 152.
8. Hahn, p. 155.
9. Quoted in Emling, p. 129.
10. Hahn, p. 160.
11. The decays of actinium reach effective stability at bismuth (atomic number 83). Bismuth has a half-life of 10^{19} years, which is a billion times longer than the age of the universe.
12. Meitner confirmed the existence of protactinium in 1917.
13. The ordering of elements by these atomic numbers came from theorist Niels Bohr and experimentalist Henry Moseley in 1913.
14. Subsequently established as protactinium (91 in 1913), francium (87 in 1939), and astatine (85 in 1940).

Chapter 4: The Nuclear Atom
1. Rutherford laboratory notebooks, PA175, 178, and 194, Cambridge University Library. This period is described in detail in Wilson, Chapter 10.
2. Rutherford anecdote quoted in Wilson, p. 291.
3. Rutherford anecdote quoted in Wilson, p. 291.
4. Darwin memoirs and Rutherford correspondence, quoted in Wilson, p. 295.
5. Talk at Manchester Literary Society, 7 March 1911, quoted in Wilson, p. 298.
6. Quoted in Spark Notes, "Niels Bohr: Biography", accessed 24 September 2024, www.sparknotes.com/biography/bohr/section3/.

Chapter 5: Rutherford "Splits the Atom"
1. Rutherford's apology to the anti-submarine warfare committee, quoted in Wilson, p. 405.
2. Rutherford collected papers, quoted in Wilson, p. 387.
3. Rutherford collected papers, quoted in Wilson, p. 394.
4. Quoted in Campbell, p. 376.
5. Rutherford's talk at the New Islington Hall, 7 February 1916, quoted in Oliphant, pp. 139–140.
6. Quoted in Oliphant, pp. 139–140.
7. E. Rutherford, "Nuclear constitution of atoms", *Proceedings of the Royal Society* vol. 97 (1920), p. 396.

8. Deuterium was discovered in 1931. Rutherford and two colleagues, Mark Oliphant and Paul Harteck, discovered tritium in 1934 by bombarding deuterium with high energy deuterons.

9. Rutherford, "Nuclear constitution of atoms", p. 396.

10. J. Chadwick in the foreword to Oliphant, p. x.

11. J. Chadwick, "Some personal notes on the search for the neutron", in *Proceedings of the 10th International Conference of the History of Science, Ithaca*, Hermann, 1964, pp. 159–162.

Chapter 6: The Mystery of Beryllium

1. Hahn describes their quest in Hahn, pp. 168–170.

2. Arthur Conan Doyle, *The Case-Book of Sherlock Holmes*, John Murray, 1927.

3. Frederick Soddy and John Cranston in Glasgow independently identified protactinium around this same time.

4. Dr Lise Meitner, "Die Muttersubstanz des Actiniums, Ein Neues Radioaktives Element von Langer Lebensdauer", *Zeitschrift für Elektrochemie und angewandte physikalische Chemie* vol. 24 (1 June 1918), p. 169.

5. In 1900 Becquerel had shown the negative charge of beta rays by detecting the direction of their deflection in a magnetic field. This would be no good for positive betas, however, because antimatter positrons annihilate with electrons in the surrounding matter almost immediately. The positron was predicted theoretically in 1931 and discovered in 1932.

6. The minutes of the committee meeting regarding the award of the Lieben Prize in physics 1925, as quoted in the English translation of *Women in European Academies*, edited by Ute Frevert, Ernst Osterkamp, and Günter Stock, De Gruyter Akademie Forschung, 2021, available at https://blog.degruyter.com/lise-meitner-a-physicist-who-never-lost-her-humanity.

7. This would be the first of nineteen nominations for Meitner in chemistry: Nobel Prize.org, "Nomination for Nobel Prize in Chemistry 1924", accessed 25 September 2024, www.nobelprize.org/nomination/archive/show.php?id=5722. She was also nominated for physics on twenty-nine occasions between 1937 and 1967: Margaret Harris, "Overlooked for the Nobel: Lise Meitner", *Physics World*, 5 October 2020, www.physicsworld.com/a/overlooked-for-the-nobel-lise-meitner/.

8. More innocently, Frédéric wanted to ensure that his and Irène's offspring would carry forward the name Curie. At the Cavendish Laboratory, however, James Chadwick always referred to them as M. Joliot and Mme Curie-Joliot.

9. F. Joliot and I. Curie, "Émission de protons de grande vitesse par les substances hydrogénées sous l'influence des rayons gamma très pénétrant", *Comptes Rendus hebdomadaries des séances de l'Académie des Sciences* vol. 194 (18 January 1932), p. 293.

Chapter 7: Il Papa

1. Segrè and Bettina, p. 3.

2. Armin Hermann, *Max Planck: The Genesis of Quantum Theory*, MIT Press, 1971, p. 74.

3. Schwartz, p. 59.

4. Fermi's biographers give no precise date for this meeting. Fermi's paper "A statistical method to determine some properties of atoms" was presented at the Accademia dei Lincei on 7 December 1927, so the meeting with Majorana was probably in October or November.

5. Segrè, p. 50.

6. Segrè and Bettina, p. 81.
7. Schwartz, p. 108.
8. Enrico Fermi, *Collected Papers*, vol. 1, edited by E. Segrè, University of Chicago Press, 1962, p. 33.
9. Pauli's letter and its English translation are available at: Stack Exchange Physics, "'Dear Radioactive Ladies and Gentlemen'—Letter by Wolfgang Pauli", 3 March 2012, https://physics.stackexchange.com/questions/21814dear-radioactive-ladies-and-gentlemen-letter-by-wolfgang-pauli.
10. Fermi as reported in Recami, p. 2.
11. Segrè, p. 50; Emilio Segrè, Interview with Thomas S. Kuhn, American Institute of Physics, 18 May 1964, www.aip.org/hisotry-programs/niels-bohr-library/oral-histories. Quoted in Esposito, p. 16.
12. Emilio Segrè, *From X-Rays to Quarks: Modern Physicists and Their Discoveries*, W. H. Freeman, 1980, p. 177.
13. Segrè and Bettina, p. 86.

Chapter 8: In Bed for a Fortnight

1. The journal was issued on 18 January in Paris. Chadwick's colleague Norman Feather gave a talk about the paper to the Cavendish Laboratory's Kapitza Club on 26 January. Folklore is that they received the journal on a Friday.
2. J. Chadwick, "Some personal notes on the search for the neutron", in *Proceedings of the Tenth Annual Congress of the History of Science*, Hermann, 1964, pp. 159–162, cited in Rhodes, *The Making of the Atomic Bomb*, p. 162.
3. It is often claimed that Rutherford—convinced they were on the right track—offered to loan his Nobel Prize medal for Chadwick to use as a target of gold. However, gold is not listed among the elemental targets in Chadwick's full paper, "The existence of a neutron", *Proceedings of the Royal Society A* vol. 136 (1 June 1932), pp. 692–708.
4. Snow, p. 85.
5. J. Chadwick, "Possible existence of a neutron", *Nature* vol. 129, no. 312 (1932), pp. 312–313.
6. A modern replication of this by Prof Valerie Gibson of Trinity College is available at: Trinity College, Cambridge, "Professor Val Gibson recreates Newton's famous speed of sound experiment in Nevile's Court", YouTube, 20 October 2016, www.youtube.com/watch?v=Gy7HqToiBvo.
7. Snow, p. 85.
8. Chadwick, "The existence of a neutron".
9. Meeting 299 on 26 January recorded the main talk by Paul Dirac on a new paper on relativistic collision theory, followed by Feather's presentation: "Experiments of Curie and Joliot, action of Be[ryllium] γ [gamma] rays on free hydrogen".
10. Article by "Our London correspondent" [J. G. Crowther], *Manchester Guardian*, 27 February 1932, quoted in Brown, p. 111.
11. Brown, p. 111.
12. "A New Ray; Dr Chadwick's Search for 'Neutrons'", *The Times*, 29 February 1932.

Chapter 9: Moonshine

1. Ernest Rutherford, "Address of the President, Sir Ernest Rutherford, O. M., at the Anniversary Meeting, November 30, 1927", *Proceedings of the Royal Society* vol. 117 (2 January 1928), p. 310.

2. George Gamow, "Zur Quantentheorie des Atomkernes" (On the quantum theory of the atomic nucleus), *Zeitschrift für Physik* vol. 51 (1928), pp. 204–212.

3. This synchrony breaks down for particles moving at significant fractions of light speed, where Einstein's relativistic mechanics take over. For this reason, cyclotrons have no place in modern high energy particle physics.

4. Lord Bowden, lecture at the University of Canterbury, New Zealand, 15 March 1979, quoted in Cathcart, p. 223.

5. Oliphant, p. 86.

6. J. Cockcroft and E. Walton, "Disintegration of lithium by swift protons", *Nature* vol. 129 (1932), p. 649.

7. This remark by biologist John D. Bernal is quoted in: Andrew Brown, "J D Bernal: The sage of science", *Journal of Physics: Conference Series* 57 (2007), p. 67.

8. E. Rutherford, "Discussion on the structure of atomic nuclei", *Proceedings of the Royal Society* A136 (1932), pp. 735–762.

9. A. F., "Atomic transmutation", *Nature* 132 (1933), p. 433. The full quote is from the Associated Press as in: "Atom-powered world absurd, scientists told", *New York Herald Tribune*, 12 September 1933. But see also page 135 and notes 11 and 12 in Chapter 10.

Chapter 10: The Magicians

1. Enrico Fermi to Edoardo Amaldi, quoted by Amaldi in a letter to Erasmo Recami, 19 July 1965: Recami, p. 2.

2. "Before Easter" is from an Amaldi interview quoted in Esposito, p. 28.

3. Amaldi, Interview with Thomas Kuhn; also Esposito, p. 28.

4. Esposito, p. 29.

5. Ettore Majorana, "Uber die Kerntheorie" (On nuclear theory), *Zeitschrift für Physik* vol. 82 (1937), p. 137.

6. Majorana, "Uber die Kerntheorie", p. 137.

7. Ettore Majorana, letter to his family, quoted in Esposito, p. 32. Heisenberg's paean to Majorana arising from their 1933 collaboration is quoted in Esposito, pp. 32–33.

8. Deuterium was discovered in 1931 by American chemist Harold Urey and tritium in 1934 by Rutherford, Oliphant, and Harteck. The concept of shells was in Majorana's first seminar in Rome, recounted by Amaldi: see Esposito, p. 28.

9. Ettore Majorana, letter to his family, quoted in Esposito, p. 32.

10. Werner Heisenberg, Interview with Thomas S. Kuhn, Session X, American Institute of Physics, 28 February 1963, www.aip.org/history-programs/niels-bohr-library/oral-histories/4661-10; also quoted in Esposito, pp. 32–33.

11. British Association for the Advancement of Science, *Report of the Annual Meeting, 1933*, Office of the British Association, 1933, available at www.biodiversitylibrary.org/item/96144#page/4/mode/1up.

12. See, for example, Quote Investigator, "Anyone who expects a source of power from the transformation of these atoms is talking moonshine", 26 November 2018, https://quote-investigator.com/2018/11/26/moonshine/.

13. Snow, p. 103.

14. Quoted in Rhodes, *The Making of the Atomic Bomb*, p. 25.
15. Knowledge of London, "Little Eve", accessed 26 September 2024, http://Knowledge oflondon.com/firsttrafficlights.html.
16. Adding a neutron to beryllium-9 forms beryllium-10 which beta decays to produce boron-10.
17. Szilard's machinations are described in Farmelo, pp. 72–78.

Interlude
1. Lise Meitner, article in Solvay, p. 176; quoted in Guerra et al., p. 9.
2. Quoted in Guerra et al., p. 12.
3. Quoted in Esposito, p. 32.
4. N. Bohr, *The Solvay Meetings and the Development of Quantum Mechanics*, section VII, October 1961, www.solvayinstitutes.be/pdf/Niels_Bohr.pdf.
5. Hans Bethe, Interview with Charles Weiner and Jagdish Mehra, Session I, American Institute of Physics, 27 October 1966, www.aip.org/history-programs/niels-bohr-library/oral-histories/4504-1.

Chapter 11: Fermi Explains Beta Radioactivity
1. Segrè, p. 89.
2. The force is today more generally known as the weak force. I included "nuclear" to differentiate it in this particular example of Fermi's creation from the strong nuclear force that binds nuclei together.
3. Segrè, p. 90.
4. Enrico Fermi, "Tentativo di una teoria dell'emissione dei raggi beta", *La Ricerca Scientifica* vol. 2, no. 12 (1933); Enrico Fermi, "Versuch einer theorie der beta-strahlen", *Zeitschrift für Physik* vol. 88 (1934), p. 161. An English translation is in Fred L. Wilson, "Fermi's theory of beta decay", *American Journal of Physics* vol. 36 (1968), pp. 1150–1160.
5. Peierls, p. 103.
6. The neutrino was discovered in 1956. Its history and role in modern physics are described in Close, *Neutrino*.
7. Fermi, *Atoms in the Family*, p. 84.

Chapter 12: Third Time Lucky
1. Lise Meitner, letter to Irène Curie, 18 November 1933, in The Papers of Lise Meitner, Churchill Archives Centre, Churchill College, Cambridge.
2. An annotation by Frédéric fifteen years later in his manuscript notes for a 1948/1949 lecture course at the Sorbonne, in Archives Irène Curie et Frédéric Joliot-Curie (1918–1958), Paris; quoted in Guerra et al., p. 13.
3. Guerra et al., p. 15.
4. I. Curie and F. Joliot, "Un nouveau type de radioactivité", *Comptes Rendus hebdomadaries des séances de l'Académie des Sciences* vol. 198 (1934), pp. 254–256; F. Joliot and I. Curie, "Artificial production of a new kind of radio-element", *Nature* vol. 133 (1934), pp. 201–202.
5. Frédéric Joliot, "Chemical evidence of the transmutation of elements", Nobel Lecture, 12 December 1935, www.nobelprize.org/uploads/2018/06/joliot-fred-lecture.pdf, p. 373.

Chapter 13: To Uranium and Beyond
1. I. Curie and F. Joliot, "Un nouveau type de radioactivité", *Comptes Rendus hebdomadaries des séances de l'Académie des Sciences* vol. 198 (1934), pp. 254–256; F. Joliot

and I. Curie, "Artificial production of a new kind of radio-element", *Nature* 133 (1934), pp. 201–202.

2. Segrè, p. 97.

3. In the periodic table, lanthanum and thorium are chemically similar. The quote is from Rhodes, *The Making of the Atomic Bomb*, p. 234.

4. Ida Noddack, "Über das Element 93" (On element 93), *Angewandte Chemie* vol. 47 (1934), pp. 653–655.

5. Emilio Segrè, Interview with Charles Weiner and Barry Richman, American Institute of Physics, 13 February 1967, www.aip.org/history-programs/niels-bohr-library/oral-histories/4876.

6. In addition, Walter was nominated nine times.

7. Segrè, p. 117.

8. Ida Noddack, "Das periodische System der Elemente und seine Lücken", *Angewandte Chemie* vol. 47 (1934), pp. 301–305. The remark about "dogma" is quoted in G. M. Santos, "A tale of oblivion: Ida Noddack and the 'universal abundance' of matter", *Notes and Records of the Royal Society London* vol. 68, no. 4 (20 December 2014), pp. 373–389.

9. Bruno Pontecorvo, unpublished biographical notes, described in Close, *Half-Life*, p. 18.

10. Fermi, *Atoms in the Family*, p. 100.

Chapter 14: Majorana's Vision

1. The semi-empirical mass formula admits the possibility that if some 10^{43} neutrons grip one another through their mutual gravitational attraction, a massive nucleus of neutrons could result. Such entities indeed exist—they were detected by astronomers and are known as neutron stars.

2. Quoted in D. B. Meli, "Cotes, Roger (1682–1716)", *Oxford Dictionary of National Biography*, 23 September 2004, https://doi.org/10.1093/ref:odnb/6386.

3. Rudrangshu Mukherjee, "Frank Ramsey: A giant among titans", *The Wire*, 30 August 2020, https://thewire.in/books/frank-ramsey-a-giant-among-titans.

4. For example, Antonino Zichichi, "Ettore Majorana: Genius and mystery", *CERN Courier* vol. 46, no. 6 (2006), p. 23.

5. The evidence is assessed forensically in Esposito, p. 159 et seq.

6. Leonardo Sciascia, *The Moro Affair and the Mystery of Majorana*, Carcanet Press, 1987.

7. Esposito, p. 149.

8. Amaldi, "Riccordo di Ettore Majorana", quoted in Esposito, p. 149.

9. Frédéric Joliot, "Chemical evidence of the transmutation of elements", Nobel Lecture, 12 December 1935, www.nobelprize.org/uploads/2018/06/joliot-fred-lecture.pdf, p. 373.

Chapter 15: A Walk in the Woods

1. Niels Bohr, "Transmutations of atomic nuclei", *Science* vol. 86 (20 August 1937), p. 161. This idealised model ignores dissipative effects of friction.

2. As recalled by his wife: Hanni Bretscher, Interview with John Bennett and Anna Shepherd, American Institute of Physics, 10 July 1984, www.aip.org/history-programs/niels-bohr-library/oral-histories/4536.

3. Bretscher, Interview.

4. M. Longair, personal email, 22 October 2023; E. Bretscher and L. G. Cook, "Transmutations of uranium and thorium nuclei by neutrons", *Nature* vol. 143 (1939), pp. 559–560.

5. Bretscher, Interview.

6. I. Curie and P. Savitch, "Sur les radioéléments formés dans l'uranium irradié par les neutrons", *Journal de Physique et le Radium* vol. 8 (October 1937), pp. 385–387.

7. O. Hahn and F. Strassman, "Concerning the existence of alkaline earth metals resulting from neutron irradiation of uranium", *Naturwissenschaften* vol. 27 (January 1939), pp. 11–15; O. Hahn and F. Strassman, "Verification of the creation of radioactive barium isotopes from uranium and thorium by neutron irradiation; Identification of additional radioactive fragments from uranium fission", *Naturwissenschaften* vol. 27 (February 1939), pp. 89–95.

8. Frisch, p. 115.

9. This energy is vast compared to that released in chemical reactions for the following reason. Energy manifested by electrical repulsion is proportional to the product of the charges involved and inversely proportional to their spatial separation. Chemical reactions involve the movement of individual electrons at atomic distance scales. Fission of uranium, by contrast, involves up to ninety-two protons packed within an atomic nucleus, some ten thousand times more compact than the size of an atom. The resulting electrostatic energy in fission, relative to chemical reactions, is thus elevated by the product of the nuclear charges and a further ten thousand for the more compact situation.

10. Frisch, p. 116.

11. Otto Frisch, "The discovery of fission: How it all began", *Physics Today* vol. 20 (November 1967), p. 48.

Chapter 16: Chain Reaction

1. Lew Kowarski, Interview with Charles Weiner, Session I, American Institute of Physics, 20 March 1969, www.aip.org/history-programs/niels-bohr-library/oral-histories/4717-1.

2. See Holloway, p. 57, and Marnham, p. 189. See also Close, *Half-Life*, Chapter 3.

3. Teller, *The Legacy of Hiroshima*, p. 9 et seq.

4. H. L. Anderson, E. Fermi, and L. Szilard, "Neutron production and absorption in uranium", *Physical Review* vol. 56 (1 August 1939), pp. 284–286, received by the editor on 3 July 1939. More details are in Schwartz, p. 163 et seq.

5. Remark by Oppenheimer's student, Philip Morrison, quoted in Rhodes, *The Making of the Atomic Bomb*, p. 274.

Chapter 17: "Extremely Powerful Bombs"

1. Siegfried Flügge, "Kann der Energieinhalt der Atomkerne technisch nutzbar gemacht werden?", *Die Naturwissenschaften* vol. 27 (9 June 1939), pp. 402–410; C. P. Snow, "A new means of destruction", *Discovery* vol. 2 (September 1939), pp. 443–444.

2. Final sentence in H. L. Anderson, E. Fermi, and L. Szilard, "Neutron production and absorption in uranium", *Physical Review* vol. 56 (1 August 1939), p. 286.

3. Albert Einstein, letter to President Roosevelt, 2 August 1939. Available at www.atomicarchive.com/resources/documents/beginnings/einstein.html.

4. Fermi, *Atoms in the Family*, p. 164.

5. Farmelo, p. 129.

6. Lew Kowarski, Interview with Charles Weiner, Session II, American Institute of Physics, 19 October 1969, www.aip.org/history-programs/niels-bohr-library/oral-histories/4717-2.

7. G. Flerov and C. Petrjak, "Spontaneous fission of uranium", *Physical Review* vol. 58 (1 July 1940), p. 89.

8. Peierls in Frank Close, "Destroyers of Worlds", BBC Radio 4, 11 July 2015, www.bbc.co.uk/programmes/b061pchg.

9. Peierls in Close, "Destroyers of Worlds".

10. Frisch–Peierls Memorandum, March 1940, File AB 1/210, UK National Archives, Kew. Quoted in full in Close, *Trinity*, pp. 26–29.

11. Peierls, p. 155.

12. Today centrifuges are used, but in 1941 the metallurgy was not sufficiently robust.

13. Oliphant, "The beginning: Chadwick and the neutron", *Bulletin of the Atomic Scientists* vol. 38 (December 1982), pp. 14–18.

14. The eventual cost has been estimated at about two billion dollars (Serber, p. 14), most of which was in building the machinery required to separate uranium-235.

Chapter 18: A Nuclear Engine

1. According to Hahn, pp. 18 and 34, this came during their attempts to replicate Fermi's 1934 claims of transuranic elements, but no date is given.

2. Gowing, p. 60.

3. Hanni Bretscher, Interview with John Bennett and Anna Shepherd, American Institute of Physics, 10 July 1984, www.aip.org/history-programs/niels-bohr-library/oral-histories/4536.

4. Frédéric Joliot-Curie's plans for a nuclear reactor ended with the fall of France in 1940.

5. Arthur Compton to James Conant, quoted in Rhodes, *The Making of the Atomic Bomb*, p. 442.

6. The origins, goals, and development of the Manhattan Project are summarised in Atomic Archive, "The Manhattan Project", accessed 25 September 2024, www.atomicarchive.com/resources/documents/manhattan-project/index.html.

7. Serber, p. xi.

8. Teller, *Memoirs*, p. 143.

9. Serber, p. xiv.

10. The words come from the Sanskrit scriptural text the Bhagavad Gita.

11. Alex Wellerstein, "Counting the dead at Hiroshima and Nagasaki", *Bulletin of the Atomic Scientists*, 4 August 2020, https://thebulletin.org/2020/08/counting-the-dead-at-hiroshima-and-nagasaki/.

12. Sime, p. 258; O. R. Frisch, "Lise Meitner 1878–1968", *Biographical Memoirs of Fellows of the Royal Society, London* vol. 16 (1970), p. 414.

13. The letter and some signatures can be found at National Archives Catalog, "Petition from Leo Szilard and Other Scientists to President Harry S. Truman", 17 July 1975, https://catalog.archives.gov/id/6250638.

14. For a full record of Fuchs's espionage as well as his invaluable contributions to development of both the atomic and later the hydrogen bomb, see Close, *Trinity*, op cit.

Chapter 19: Destroyers of Worlds
  1. Teller, *The Legacy of Hiroshima*, p. 37.
  2. Teller, *Memoirs*, p. 159.
  3. It was to replace Teller in this role that Peierls came to Los Alamos.
  4. Fermi, *Super Lectures*, Lecture 6.

Chapter 20: The Ulam–Teller Invention
  1. The method is named after the Monte Carlo Casino because the technique uses random number generators to simulate possible outcomes of an uncertain event. The different outcomes mimic real-life situations, which are then assessed using probability theory.
  2. On 31 October at 7:15 p.m. GMT, but on 1 November local time in Elugelab.
  3. Quoted in WGBH, "U.S. Tests", accessed 25 September 2024, www.pbs.org/wgbh/americanexperience/features/bomb-us-tests/.
  4. Serber, p. 4, fn. 2.
  5. Fuchs's story is in Close, *Trinity*.
  6. The Kuiper Belt is a region of icy objects in the solar system extending beyond the orbits of Neptune and Pluto.
  7. Goncharov, "American and Soviet H-bomb development programmes", p. 1036.
  8. Sakharov, p. 92.
  9. And possibly also from physicist Ted Hall, who was at Los Alamos. Hall's story is in *Bombshell* by Albright and Kunstel.
  10. Goncharov, "On the history of creation of the Soviet hydrogen bomb", p. 861, gives 23 April 1948 as the date when Beria took action on information passed by Fuchs on 13 March about the Fuchs–von Neumann invention. A full chronology of Fuchs's espionage is in Close, *Trinity*.
  11. Goncharov, "American and Soviet H-bomb development programmes", p. 1039. Sakharov remarked: "I gave it no thought at the time, but I now believe that the design developed by the Zeldovich group for a hydrogen bomb was directly inspired by information acquired through espionage" (Sakharov, p. 94).
  12. Sakharov, p. 102.
  13. Goncharov, "American and Soviet H-bomb development programmes", p. 1038.
  14. R. F. Taschek et al., "A study of the interaction of protons with tritium", *Physical Review* vol. 75 (1 May 1949), pp. 1361–1365.
  15. Sakharov, p. 174.
  16. Sakharov, p. 174.
  17. Sakharov, p. 181.
  18. Sakharov, Chapter 12, "The Third Idea".
  19. Sakharov, p. 182.
  20. Sakharov, p. 225.
  21. The British told Hans Bethe in 1960 how they discovered the Teller–Ulam secret. Bethe interview on 5 January 1989, in Hirsch and Mathews, p. 437.
  22. Close, *Trinity*, pp. 389–390.
  23. Ward, p. 14; quoted in Dombey.
  24. Quoted in Dombey.

25. Author interview with Bryan Taylor, 6 May 2017, and remarks in *Britain's Nuclear Bomb: The Inside Story*, BBC Radio 4, 3 May 2017.

26. Author interview with Lorna Arnold, official historian of the British H-bomb project, 30 August 2013, and Close, *Trinity*, p. 391.

Chapter 21: The MADness of Tsar Bomba

1. Quoted in Viktor Adamsky and Yuri Smirnov, "Moscow's biggest bomb: The 50-megaton test of October 1961", *Cold War International History Project Bulletin* vol. 4 (Fall 1994), p. 20, www.wilsoncenter.org/sites/default/files/media/documents/publication/CWIHP_Bulletin_4.pdf.

2. Sakharov, p. 217.

3. Sakharov, p. 220.

4. Vladimir Afanasyev, quoted in Vladimir Suvorov, *Strana Limoniya*, Soviet Russia Press, 1989, pp. 117–127, as cited by Adamsky and Smirnov, "Moscow's biggest bomb", p. 19.

5. Adamsky and Smirnov, "Moscow's biggest bomb", p. 19.

6. Sakharov, p. 221.

7. Frisch–Peierls Memorandum, March 1940, File AB 1/210, UK National Archives, Kew. Quoted in Close, *Trinity*, p. 26.

8. For an illustration of the enormity of 50 megatons relative to 20 kilotons, see Alex Wellerstein, "An unearthly spectacle: The untold story of the world's biggest nuclear bomb", *Bulletin of the Atomic Scientists*, 29 October 2021, https://thebulletin.org/2021/11/the-untold-story-of-the-worlds-biggest-nuclear-bomb/.

9. I. I. Rabi, quoted in Alex Wellerstein, "In search of a bigger boom", *Restricted Data: A Nuclear History Blog*, 12 September 2012, https://blog.nuclearsecrecy.com/2012/09/12/in-search-of-a-bigger-boom/.

10. Its energy was less than Krakatoa, but its power greater.

11. See also Mercurio 2007.

Postscript

1. NobelPrize.org, "The Nobel Prize in Chemistry 1944", accessed 8 October 2024, www.nobelprize.org/prizes/chemistry/1944/summary/.

2. NobelPrize.org, "The Nobel Prize in Chemistry 1944".

3. Operation "Epsilon" (6–7 August 1945), National Archives and Records Administration, College Park, MD, RG 77, Entry 22, Box 164 (Farm Hall Transcripts); available at German History in Documents and Images, "Transcript of surreptitiously taped conversations among German nuclear physicists at Farm Hall (August 6–7, 1945)", September 5, 2024, https://germanhistorydocs.org/en/nazi-germany-1933-1945/transcript-of-surreptitiously-taped-conversations-among-german-nuclear-physicists-at-farm-hall-august-6-7-1945.

4. Joseph Rotblat, "Leaving the bomb project", *Atomic Heritage Foundation*, accessed 25 September 2024, https://ahf.nuclearmuseum.org/ahf/key-documents/rotblat-account/.

5. NobelPrize.org, "The Nobel Peace Prize 1995", accessed 19 September 2024, www.nobelprize.org/prizes/peace/1995/summary/.

6. NobelPrize.org, "The Nobel Peace Prize 1975", accessed 19 September 2024, www.nobelprize.org/prizes/peace/1975/summary/.

7. NobelPrize.org, "Andrei Sakharov: Acceptance Speech", accessed 19 September 2024, www.nobelprize.org/prizes/peace/1975/sakharov-acceptance-speech/.

INDEX

Abelson, Philip, 240
actinium, 48, 49, 52, 81–85
Alarm Clock, 274, 276, 284, 290
Aldermaston, 288–290
Allier, Jacques, 227–228
alpha decay, 51 (fig.), 85, 149–150, 153
alpha particles
  beryllium bombardment, 104–106, 168
  Cockcroft and Walton, 121–124
  counting, 56–57
  energies of, 123
  Gamow, 115, 150
  Hahn, 44–45, 84
  link to helium, 50, 54, 62
  Marsden, 57–59, 71–73
  Meitner, 84
  polonium, 87–89, 107, 164
  positive charge, 58
  range, 72
  Rutherford, 30, 44–45, 55–60, 70–76, 121
  from splitting lithium, 137
alpha radiation/radioactivity
  atomic number change, 82
  Cockcroft, 115, 122
  Hahn, 44–45
  Rutherford, 30, 32, 44
  Soddy, 49–50
  transmutation by, 50
aluminium, inducing radioactivity in, 163–165, 165 (fig.), 169
Amaldi, Edoardo
  Fermi's theory, 155
  inducing radioactivity, 169
  on Majorana, 188
  photograph, 181 (fig.)
  Rome laboratory, 96
  slow neutrons, 178–180

American Castle test series, 293
Anderson, Carl, 143–144
Anderson, Herb, 213
Anglo-Canadian Collaboration, 228
anode, 13, 29
antimatter, 83, 143
argon, 36–37
Arnold, William, 202
artificial radioactivity, 162, 171
The Ascent of Man (TV series), 255
atomic bomb
  assembly, 254
  hydrogen bomb compared, 261–262
  Oppenheimer, 217–218
  Soviet Union's quest for, 260
  term coinage by Wells, 41
atomic energy
  Rutherford, 32–33
  Soddy, 39–41
atomic mass, 41
atomic mass unit, 12
atomic nucleus
  alpha particle deflection, 58–59
  Bohr, 61–63
  energy in, 75, 123, 200–201
  energy release from, 247, 257
  H-particles, 70–73
  Majorana, 128
  positive charge, 58–61, 63
  as radioactivity origin, 60–61
  relative proportion of neutrons and protons, 183–186
  Rutherford discovery, 58–60
  Solvay Conference (1933), 142–146
atomic number, 62, 70, 82, 177
atom smasher, 122

Index

Bakerian Medal and Lecture, 76–78, 80, 90, 101, 109
Balmer, Johann, 64–65
barium, 194–201
   Bretscher, 194–195
   coalescence with krypton, 200
   Hahn, 194–198
   isotopes of, 196
   from splitting uranium, 196, 199, 201
Becker, Herbert, 88–89, 104
Becquerel, Henri
   afterword on, 311
   phosphorescence, 16–17, 19
   radiation, 20, 28, 53
   uranium experiments, 17–20, 18 (fig.)
Berg, Otto, 175
Beria, Lavrentiy, 282–284
beryllium
   alpha particle bombardment, 101, 104–108, 168
   splitting, 138, 138 (fig.)
beta decay
   conversion to even-even state, 184
   Fermi, 150, 152, 154–158, 162, 167, 214
   Heisenberg's uncertainty principle, 151
   Joliot-Curies, 164
   Majorana, 184–185
   Pauli, 99–100, 149, 152, 154
   radioactive ladders, 51 (fig.), 52, 82–84, 99–100, 143–144
   Solvay Conference (1933), 143–145, 149–150, 152
beta particle, 62
beta radiation/radioactivity
   atomic number change, 82
   Fermi explanation, 149–159
   Hahn and Meitner, 52
   Rutherford, 30
Bethe, Hans
   afterword on, 314
   departure from Germany, 157
   at Los Alamos, 256
   neutrino, 158–159, 214
   on Solvay Conference, 146
   Teller and, 263–265, 267
   uranium capturing neutrons, 192
Big Bang, 122
big science, 119
Bikini Atoll, 281
binding energy, 71, 128–129, 156, 185, 199
bismuth, 197
black body radiation, 278
Bohr, Niels, 60–66
   abandonment of energy conservation in the nucleus, 152
   afterword on, 312

   on atomic bomb production, 250
   on atomic explosion, 224–225, 229
   atomic model, 94–95, 129
   atomic number, 70, 82
   atomic structure, 54, 61–63
   Fermi's escape from Italy, 196
   fission, 202–204
   Frisch, 191–193, 201–202
   on isotopes, 62
   liquid drop analogy, 193, 198–199
   meeting Rutherford, 60
   periodic table explanation, 61–62
   quantum theory, 94
   Solvay Conference (1933), 145
   why fission of natural uranium is not explosive, 214–219
Boltwood, Bertram, 43–44, 47, 59
Bonner, Elena, 310
boosted fission bomb, 270–271
boron
   inducing radioactivity in, 163–165, 165 (fig.)
   splitting of, 124
Bothe, Walher, 88–89, 104
Bretscher, Egon, 193–195, 240
Briggs, Lyman, 237
Bronowski, Jacob, 255, 269, 271

cadmium, 245, 247
Cairncross, John, 259
calorie, 32
Campbell, John, 27
Capon, Laura, 93
Castle Bravo test, 281–282, 286, 289
cathode, 13, 29
cathode rays, 15, 22–23, 29–30
centrifuge, 224–225, 234, 251
Chadwick, James
   afterword on, 313
   early career, 79
   neutrons, 80, 103–112, 111 (fig.), 120, 128, 143, 158, 166
   report to Churchill, 235–236
   Rotblat and, 234–235, 308
chain reaction, 211–218
   Fermi, 225, 242–246
   Joliot, Frédéric, 209–211, 226
   Kurchatov, 229
   Majorana, 189–190
   nuclear fission, 223–225
   nuclear reactor, 226–228
   in plutonium, 248
   potential implications, 211–212
   Szilard, 136–139, 189, 224
   Trinity test explosion, 3–4
   why Earth's crust isn't exploding, 214–218

Index

charge-to-mass ratio, 23
Chernobyl, 247, 301
Cherwell, Lord, 235, 242, 289
chlorine gas, 49
Churchill, Winston, 235–237, 249, 289
classical mechanics, 94–95, 115
Classical Super bomb, 266, 268–271, 276, 283, 285, 302
cloud chamber
   Anderson, 143
   Chadwick and Webster, 105
   Cockcroft and Walton, 123–124
   Dee, 135
   Frisch, 201–203
   Hahn and Meitner, 83–84
   Joliot Curies, 143, 162
   Rasetti, 99
Cockcroft, John, 115–117, 120–124, 134, 137, 313
collaborations, 170, 228
colours, 64
compression
   in hydrogen bomb, 281
   ionisation, 270–272, 284
   radiation, 271–272, 287
   supercompression, 276–278
   symmetrical, 285
computer science, and von Neumann, 269–271, 275–276, 278
Cook, Leslie, 193–195
Corbino, Orso
   early life and career, 92
   Fermi and, 17, 92–93, 95, 98, 151, 171
   home of, 96
   as Padreterno, 98
   patent advice, 180
   recruitment of students to physics, 96
   Rome conference, 151
Corelli, Antonio, 188
Cotes, Roger, 186
COVID pandemic, 34
critical mass, 3–4, 236, 256
critical size, 231–232, 254, 264
Crookes, William, 14–15, 82
Crowther, J. G., 112
Curie (unit), 22
Curie, Irène, 85–87. See also Joliot-Curie, Irène
Curie, Marie, 20–22
   death, 171
   early life, 19–20
   illness, 21, 86
   Nobel Prize, 22
   radioactivity term coined by, 19
   radium, 19, 21, 30, 35, 42
   Radium Institute, 85–87
   Rutherford's meeting, 37–38
   Solvay Conference (1933), 142

Curie, Pierre, 20–22
   afterword on, 311
   death, 85
   Nobel Prize, 22
   radium, 19, 21, 30, 35, 42
   Rutherford's meeting, 37–38
current, electric, 7, 13, 24
cyclotron
   Frédéric Joliot, 171
   Lawrence, 99, 119–120, 236, 248
   Szilard, 136

D'Agostino, Oscar, 169, 171, 181 (fig.)
Darwin, Charles, 59
Debierne, André-Louis, 48, 87
de Broglie, Louis, 65
decay. See alpha decay; beta decay
Dee, Philip, 135
Dempster, Arthur, 216
deuterium
   heavy hydrogen, 78, 133
   in hydrogen bomb, 262–272, 274–276, 279, 282–284, 290, 302
   tritium fusion with, 267, 267 (fig.), 270, 302
deuteron, 78, 158, 266, 267 (fig.)
disintegration
   by Cockcroft and Walton, 124–125
   energy liberated by, 124–125
ectoplasm, 14
Einstein, Albert
   $E = mc^2$ equation, 41, 99–100, 201
   equivalence of energy and mass, 41, 123, 153, 266
   leaving Germany (1933), 142
   letter to Roosevelt, 224
   Nobel Prize, 64
   photons, 64
   relativity, 41
   Szilard, 136, 224
   on von Neumann, 255
eka-tantalum, 12, 49
elastic scattering, 214
electric charge, 22–23
electric current, 7, 13, 24
electricity
   Faraday, 26
   from nuclear energy, 247, 299
   "opposite," 60
   properties of, 7, 13–14
   Thomson experiments, 22–24, 29
   what is electricity, 22–24
electric motors, 6
electrolysis, 13–14
electromagnetic force, 116, 155–156
electromagnetic separator, 236

Index

electromagnetic waves, 27
electromagnetism, Maxwell's theory of, 13
electrometer, 29, 33–34
electron cloud, 95, 97
electrons
  as beta particles, 54
  Bohr, 63–65, 129
  in cathode rays, 29–30
  discovery, 7, 23–24, 28, 31, 36, 54
  effects of, 8
  negative charge, 63
  quantum theory, 63, 65
  Rutherford, 77–80
  wavelike character, 65
electrostatic energy, 186, 200–201
Elugelab, 279
emanation, 34–35, 37
$E = mc^2$ equation, 41, 99–100, 123–124, 128, 185, 201
energy. *See also* nuclear energy
  of atomic nucleus, 75, 123, 200–201, 247, 257
  binding, 71, 128–129, 156, 185, 199
  calorie, 32
  electrostatic, 186, 200–201
  equivalence of energy and mass, 41, 123, 153, 266
  frozen, 41
  kinetic energy of heat, 11
  photons, 64–65
  rate of release, 40
  Rutherford's measurement of, 32–33
  from splitting uranium, 200–201
energy conservation, principle of, 31, 114–115, 152
Enewetak Atoll, 279279
ENIAC, 269–270, 276, 278
enrichment, 218. *See also* uranium-235, enrichment
Esposito, Salvatore, 187
ether, 13
exclusion principle, 95, 129, 132–133

Fajans, Kasimir, 82
fallout, 280–281, 287–289, 296, 300–301, 309
Faraday, Michael, 6, 26, 37
Feather, Norman, 107, 110, 112, 194, 241
Feklisov, Alexandr, 283
Fermi, Enrico
  afterword on, 314
  atomic bomb likelihood, 224
  at atomic bomb test (1945), 91–92
  beta decay, 150, 152–158, 162, 167, 214
  chain reaction, 225, 242–246
  early life, 92–93
  electron cloud, 95, 97
  fission, 212–213
  hydrogen bomb, 260, 263–264, 266–270

induced radioactivity, 167–170, 178
leaving Italy, 196
Los Alamos lectures, 260, 266, 269–270, 274, 302
neutrons, 153–154
news of nuclear fission, 204–205
Nobel Prize, 196, 204, 305
nuclear reactor, 242–247
photograph, 181 (fig.)
ranking of physicists, 127
Rome laboratory, 95–102
slow neutrons, 178–180, 192
Solvay Conference (1933), 153–154
Teller and, 262–264, 269
transuranic elements, 171–173, 177, 196, 213, 239
uranium irradiation, 171
weak nuclear force, 156
Fermi, Laura, 179–180
Fermi gas, 94
Feynman, Richard, 250
First Industrial Revolution, 6, 11–12, 22, 30
First World War, 49, 62, 79, 81, 85, 92, 311
Fischer, Emil, 45–47
fission. *See also* nuclear fission
  Bohr, 202–204
  Fermi, 212–213
  Frisch and Meitner, 197–204, 226
  Hahn, 305
  heat from, 247
  Joliot, Frédéric, 209–211
  neutrons from uranium fission, 209–211, 213–215
  Noddack, 173, 177
  odds of, 218–219
  Soviet Union experiments, 229–230
  spontaneous, 229, 253–255
  Szilard, 212–213
  Trinity test explosion (1945), 3–4
  uranium, 197–105, 208–209
fission bomb, as primary of fusion bomb, 274–278, 290
Flerov, Georgii, 229, 259–260
Floating University, 20
Flügge, Siegfried, 223
Frenkel, Yakov, 281
Frisch, Otto
  afterword on, 313
  Bohr, 191–193
  explosive liberation of nuclear energy, 190
  fission, 197–204, 226
  leaving Germany, 142
  Peierls and, 230–233, 298
  radioactive super bomb, 233
  splitting thorium, 216
  uranium enrichment, 231

336

Index

Fuchs, Klaus
  boosted fission bomb, 271–272
  at Los Alamos, 256, 280
  radiation implosion, 276, 280, 289, 291
  spy, 259–260, 280, 282–284, 289, 291
  UK bomb research, 289, 291
Fuchs–von Neumann device, 272, 291
Fukushima disaster, 301
fusion
  deuterium with tritium, 267, 267 (fig.), 270, 302
  energy source, 302
  of hydrogen nuclei, 262
  Soviet Union fusion bomb, 280–288
  sun, 262–264, 302
  X-rays in fusion bomb, 277–279

Gagarin, Yuri, 294–295
gallium, 253
gamma rays
  Bothe and Becker, 88–89
  Chadwick, 105, 108
  from cooling nucleus, 193
  as high energy particles of light, 54
  Irène Joliot-Curie, 90, 101
  Meitner, 84, 88
  Rutherford, 33
  Trinity test blast, 5
Gamow, George, 115, 150
Geiger, Hans, 56–57, 59, 312
Geiger counter, 56–57, 88–89, 162–163, 168–169
Gentner, Wolfgang, 163
Ginzburg, Vitaly, 283–285, 315
Gnomon, 304
Göhring, Oswald, 82
graphite, 106, 225, 227, 242–245, 253
Groves, Leslie, 249–251, 253, 259, 308
Gryn, Tola, 234–235

Hahn, Otto, 42–52
  afterword on, 312
  barium, 194–198
  beta radioactivity, 52
  Cook and, 193–195
  criticism of Noddack, 173–174
  during First World War, 49
  fission, 305
  Meitner collaboration, 42, 47–49, 52, 81–85, 172–174, 195–197
  Nobel Prize, 306–307
  in Ramsay's lab, 42–43
  thorium isotopes, 43–47
Halban, Hans, 210, 228
half-life, Rutherford discovery of, 34–35

Hanford, Washington, 251–253, 257
Harwell, 288–289, 299–300, 302–303, 313
heat
  in atomic nucleus, 193
  from fission, 247
heavy hydrogen, 78, 133
heavy water, 227–228
Heisenberg, Werner
  Majorana's exchange with, 130–134
  neutrons, 131–134, 154
  Solvay Conference (1933), 145–146
  uncertainty principle, 131, 133, 151, 153
helium
  alpha particle link to, 50, 54, 62
  alpha radioactivity, 44–45
Hertz, Heinrich, 13, 27–28
Hiroshima, 257–258, 267, 282
Hitler, Adolf, 191, 195, 212, 230, 232–234
H-particles, 70–74
hydrogen. *See also* deuterium; tritium
  fusion, 266
  heavy, 78, 133
  isotopes, 78
  spectrum, 64
hydrogen bomb
  atomic bomb compared, 261–262
  as destroyer of worlds, 279
  Fermi, 260, 263–264, 266–270
  moratorium on tests, 294
  Operation Greenhouse George test (1951), 272
  power of, 261–262, 264
  Teller, 262–267, 269–271, 274–277, 279–281, 284, 287–291, 303–304, 314
  Ulam-Teller invention, 277, 279–281, 287–291
  von Neumann, 269–272
hydrophones, 70

induced radioactivity
  Fermi, 167–170, 178
  Joliot-Curies (Irène and Frédéric), 162–166
  McMillan, 240
  slow neutrons, 179
Industrial Revolutions, 6–7
inert gases, 36–37, 42
Interpretation of Radium (Soddy), 40
The Invisible Man (Wells), 106
ionisation
  Chadwick experiments, 107
  detection/measurement, 29
  Rutherford, 29, 32, 33
ionisation chamber, 107–108, 203, 212
ionisation compression, 270–272, 284
ionium, 47
isomers, 207–208

Index

isotopes
  atomic mass, 128
  Majorana, 130, 133–134
  Rutherford, 78
  separation by magnetic field, 216–217
  Soddy, 49–50, 52, 62
  stability, 170
  term introduced by Soddy, 42
  thorium, 43–47
Ivy Mike test explosion, 279–280, 282

JET (Joint European Torus), 303
Joliot, Frédéric
  afterword on, 312–313
  chain reaction, 209–211, 226
  cyclotron, 171
  early years, 86–87
  on explosive transmutation, 166
  fission, 209–211
  isomers, 207–208
  name change to Joliot-Curie, 87
  nuclear reactor, 226–228
  Solvay Conference (1933), 143–144, 154, 161
Joliot-Curie, Irène, 85–87
  afterword on, 312
  birth, 20
  lanthanum, 172–173, 196, 199, 208
  Meitner letter to, 161–163
  neutron, 146
  Solvay Conference (1933), 142–146, 154, 161
  splitting uranium, 208–209
  uranium irradiation, 171–172
Joliot-Curies (Irène and Frédéric)
  beryllium, 89
  cloud chamber, 143, 162
  induced radioactivity, 162–166
  neutrons, 90, 143
  Nobel Prize (1935), 166
  photons, 104–106, 108
  positrons, 143–145, 154, 161

Kapitza, Peter, 109–110
Kapitza Club, 109–112, 111 (fig.)
Kemmer, Nicholas, 241
Kennedy, John F., 294
Keynes, John Maynard, 186–187
Khariton, Yulii, 230, 232, 282–284, 315
Khrushchev, Nikita, 294–295, 303, 309
kinetic energy of heat, 11
Kowarski, Lew, 209–210, `228
krypton, 199–201
Kurchatov, Igor
  afterword on, 314–315
  fission experiments, 209, 229, 259
  isomers, 207–208

  Joliot and, 207–209, 229
  Soviet atomic bomb project, 260, 281–283

Langevin, Paul, 38, 85–87
lanthanum, 172–173, 196, 199, 208
Lawrence, Ernest
  cyclotron, 99, 119–120, 236, 248
  electromagnetic separator, 236
  particle acceleration, 117–120
Layer Cake, 284–285, 290
lead
  heaviest stable element, 197
  Röntgen's use of, 15–16
  uranium decay to, 44, 171
light
  properties, 64–65
  from Trinity test explosion (1945), 4–5
Lindemann, Frederick, 235–237
lithium, splitting of, 121–123, 134–138, 138 (fig.)
lithium-6, 275, 275 (fig.), 284, 290
lithium deuteride, 274–278, 280–281, 284–285, 290
Livingston, M. Stanley, 119
Lord Kelvin, 22, 24, 28
Los Alamos
  Fermi lectures, 260, 266, 269–270, 274, 302
  nerve center of Manhattan Project, 250–251, 254
  Peierls at, 256
  scientific team members at, 256, 258, 265, 280, 288–289
  Trinity test (1945), 1–2, 257
  Ulam at, 273–274
  von Neumann at, 255–257, 269–275

Maclaurin, James, 27
magnetism, and electrons, 94–95
Majorana, Ettore
  atomic nucleus theory, 128
  chain reaction, 189–190
  disappearance, 186–190
  Heisenberg exchange with, 130–134
  joining Fermi, 97–98
  neutrons, 100–102, 104, 153–154, 167–168
  nuclear forces, theory of, 130–132, 134, 145, 187
  quantum mechanics, 128–129
  stability of nuclei, 183–186
  theory of nuclear structure, 217
Malenkov, Georgy, 286
Malyshev, Vyacheslav, 286
Manhattan Project. *See also specific individuals*
  British scientists, 249, 288, 299, 313
  Einstein's letter, 224
  Feynman, 250
  Los Alamos as nerve center of, 250–251, 254

Index

Meitner's refuse to join, 258
Oppenheimer as scientific director, 250
Rabi, 265
Rotblat departure from, 307–308
shake as unit of measurement, 268
spying by Fuchs, 259
start of, 249
MANIAC, 269, 278–279, 285
Marconi, Guglielmo, 26, 28
Marsden, Ernest
  afterword on, 312
  alpha particles, 57–59, 71–72
  H-particles, 71–73
mass formula, semi-empirical, 186, 189, 200
masurium, 175–177
Maxwell, James Clark, 13, 156
McMahon Act, US, 288, 299
McMillan, Edwin, 239–240, 239–241
mechanics
  classical, 94–95
  Newton's laws of, 63
  quantum (*see* quantum mechanics)
Meitner, Lise, 47–49
  afterword on, 312
  atomic bomb, 258
  beta radioactivity, 52
  energy of beta particle, 152
  explosive liberation of nuclear energy, 190
  during First World War, 49
  fission, 197–204, 226
  Hahn collaboration, 42, 47–49, 52, 81–85, 172–174, 195–197
  leaving Germany, 195
  letter to Irène Joliot-Curies, 161–163
  omission from Nobel Prize, 306
  Solvay Conference (1933), 142–145, 152–153
  transuranic elements, 193–194
meltdown, 247, 301
Mendeleev, Dmitri, 12
mesothorium, 45–46, 49
Monte Carlo method, 273
Moseley, Henry, 61–62
mushroom cloud, 5–6, 279, 286, 296, 304–309
Mussolini, Benito, 93, 98–99, 150–151, 187

Nagasaki, 257–258, 269, 279
neptunium, 177, 215, 219, 240–242
neutral radiation, 90, 107, 109
neutrino
  Bethe and Peierls, 158–159
  Fermi, 100, 154, 156–157, 159
  Pauli, 100, 149, 152–154, 156, 158
neutrons
  capture by uranium, 219
  Chadwick, 80, 103–112, 111 (fig.), 120, 128, 143, 158, 166

chain reactions, 137–138
  fast, 216, 219, 229, 236, 241, 244, 248, 268, 270, 274
  Fermi, 153–154
  in fission process, 3–4
  Heisenberg, 131–134, 154
  to induce radioactivity, 167–169
  Joliot-Curies, 90, 143, 208
  Majorana, 100–102, 104, 128–134, 153–154, 167–168
  radioactivity induced by bombardment, 168–170
  Rutherford, 78–80, 109–110, 151
  secondary, 137–138, 209, 212–213, 225, 227, 243–244
  slow, 178–180, 192–193, 210, 216, 219, 225, 236
  Solvay Conference (1933), 143–146
  thermal, 179
  from uranium fission, 209–211, 213–215
Newton, Isaac
  classical mechanics, 63, 94–95, 115
  on Cotes, 186
  motion, laws of, 33, 119
  Principia, 186
  speed of sound, 110
nitrogen
  radioactive, 164–165
  Rutherford's bombardment of, 73–76
  transmutation to hydrogen, 76
  transmutation to oxygen, 74, 75 (fig.)
Noddack, Ida, 172–177
Noddack, Walter, 174–177
Nordmann, Charles, 76–77
nuclear atom, 53–56. *See also* atomic nucleus
nuclear electrons, 150–152
nuclear energy
  Becquerel, 19
  Bohr, 71
  early evidence of, 3, 8
  Fermi, 159
  first hint, vii
  liberation by Joliot-Curies, 166
  radioactivity as, 52
  Rutherford, 32, 75
nuclear fission
  atomic bomb, 223–238
  chain reaction, 223–225
  Frisch coinage of term, 202
  Rosenfeld reporting of, 204
nuclear force(s)
  Frisch and Meitner, 199–200
  Majorana's theory, 130–132, 134, 145, 187
  weak, 156
nuclear physics, birth of, 141–146
nuclear pile, 244–245, 247

Index

nuclear power, peaceful use of, 136, 211, 299–303
nuclear reactor
 chain reaction, 226–228
 Fermi, 242–247
 Hanford, Washington, 252–253, 257
 Joliot, 226–228
 Kurchatov, 314–315
 for plutonium synthesis, 3, 242, 248–249, 252–253, 257
 power production for peaceful use, 299–302
 theoretical in 1939, 218
nuclear stability, 183–186
nuclear test ban treaty, 303, 308, 310
nucleon, 128–129, 184–185, 192, 217–218
nucleus. *See* atomic nucleus

Oak Ridge, Tennessee, 251–252
Oliphant, Mark, 135, 237
Operation Greenhouse George test (1951), 272, 277
Oppenheimer, J. Robert
 afterword on, 314
 atomic bomb, 217–218
 Manhattan Project scientific director, 250
 Teller and, 265–266
 Trinity Test (1945), 257
oscillograph, 107

particle acceleration, 117–120
Pauli, Wolfgang
 electron spin, 94–95
 exclusion principle, 95, 129, 132–133
 neutrino, 100, 149, 152–154, 156, 158
Peierls, Rudolf
 afterword on, 313
 critical size, 231–232
 departure from Germany, 142, 157, 195
 Frisch and, 230–233, 298
 interval between successive fissions, 268
 at Los Alamos, 256
 neutrino, 158–159, 214
 radioactive super bomb, 233
Penney, William, 288–291
periodic table
 Bohr's explanation, 61–62
 gaps in, 50, 177
 Mendeleev, 12
Perrin, Francis, 144–145, 162, 164
Petrjak, Konstantin, 229
phosphorescence, 16–17, 19
phosphorus-30, 145
photo-effect, 158
photons
 detection, 88–89
 Dirac, 156

Einstein, 64
energies/frequencies, 64
isomers and, 208
Joliot-Curies, 104–106, 108
X-rays, 61
piezoelectricity, 20
pitchblende, 21, 48, 82–84
Planck, Max, 94
plutonium
 alloy, 253
 from beta decay of neptunium, 215, 219
 breeding of, 248
 Bretscher and Feather, 240–241
 name source, 241
 reactors, 3, 242, 248–249, 252–253, 257
 spontaneous fissions, 253–255
 Trinity test explosion, 2–4
 UK production of weapons-grade, 300–302
plutonium bomb, 255, 257, 269, 289
polonium
 alpha particle source, 87–89, 107, 164
 Chadwick, 107
 discovery of, 21
 Joliot-Curies, 87–88
 origin of name, 21
 radioactivity, 30, 35
 radon capsules as source, 87
Pontecorvo, Bruno, 178–179, 207–208
positron
 antimatter, 83, 143
 decay by emission of, 145
 discovery by Anderson, 143–144
 Joliot-Curies, 143–145, 154, 161
Principia (Newton), 186
protactinium, 48, 82–84
proton acceleration, 115–117, 123
protons
 fermions, 132
 Rutherford, 74, 77–80
Pugwash Conference, 308

quantum mechanics
 electron behavior, 63, 115
 Fermi, 93
 Gamow's application to nuclear physics, 115, 122
 Majorana, 128–129
 proton acceleration, 115–116
 Solvay Conference (1927), 142
quantum theory
 birth of, 94
 de Broglie, 65
 electromagnetic field, 155–156, 289
 electrons, 63–65
 Fermi, 92–95

Index

Rabi, Isidor, 170, 265, 303–304
Rabinovich, Evsei, 295
radiation. *See also* radioactivity
   alpha (*see* alpha radiation/radioactivity)
   Becquerel, 19–20, 28, 53
   beta (*see* beta radiation/radioactivity)
   Curie, 19–22
   neutral, 90, 107, 109
radiation compression, 271–272, 287
radiation exposure
   Fermi, 314
   Irène Curie, 228
   Marie Curie, 21, 171
   radium jaw, 86
   Trinity blast (1945), 5
   von Neumann, 314
radiation implosion, 276, 280, 289, 291
radioactivity
   artificial, 162, 171
   Becquerel's experiments, 19
   energy in, 71
   induced (*see* induced radioactivity)
   Rutherford, 28–38, 65
   Soddy, 36–37, 39–42, 49–50
   term invented by Curie, 19
radiochemistry, 21
radiothorium, 43, 46, 49
radiowaves, 13, 26–28
radium
   energy from, 40
   Hahn and Meitner, 47–48
   Hahn experiments, 42–43
   luminescence, 38, 86
   Marie and Pierre Curie, 19, 21, 30, 35, 42
Radium Institute, 85–87, 163, 171, 208
radon, 37
radon capsules as polonium source, 87
Ramsay, William
   Hahn and, 42–43
   inert gases, 36, 42
Ramsey, Frank, 186–187
Rasetti, Franco
   cloud chamber, 99
   Fermi's theory, 155
   inducing radioactivity, 169
   photograph, 181 (fig.)
   Rome laboratory, 96, 98
   slow neutrons, 178–179
relativity
   Einstein, 41
   Fermi, 92–93
rhenium, 175, 177
Roberts, Keith, 290
rodium, 212–213
Rome Conference (1931), 151–152
Röntgen, Wilhelm, 15–16, 19, 308, 311

Roosevelt, Franklin, 224, 236, 249
Rosenfeld, Léon, 202–204
Rotblat, Joseph
   afterword on, 315
   Chadwick and, 234–235, 308
   departure from Manhattan Project, 307–308
   Nobel Peace Prize (1995), 308
   quitting the bomb project, 258
Rutherford, Ernest, 25–38
   accomplishments, 25–26
   afterword on, 311
   alpha particles, 30, 44–45, 55–60, 70–76, 121
   ambition, 113–117
   atomic nucleus discovery, 52–53
   Bakerian Lecture (1920), 77–78, 80, 90, 101, 109
   Bohr and, 60
   on Cockcroft and Walton, 120–125
   Curie and, 85
   death, 211
   early life, 26
   gamma rays, 33
   Hahn and, 43–45
   half-life, discovery of, 34–35
   meeting the Curies, 37–38
   moonshine remark, 125, 134–135, 166
   neutral atom, 76–80
   radioactivity, 28–38, 65
   radio waves, 26–28
   Scholarship to Cambridge University, 25–27
   Soddy and, 50
   sonar, 69–70
   splitting the atom, 69–80
   Thomson and, 24, 27
   Zilard meeting (1934), 138–139

Sakharov, Andrei Dmitrievich
   afterword on, 315
   fallout implications, 288
   on John Ward, 289
   Nobel Peace Prize (1975), 309–310
   Sloika, 284
   Soviet atomic bomb project, 282–288, 291
   test ban treaty, 303
   tokamak, 303
   Tsar Bomba, 293–295, 297
Sakharov Prize, 310
Savic, Pavle, 172–173, 196, 208
Sciascia, Leonardo, 187–188
scintillation, 56–57, 72–74, 121, 123
Seaborg, Glenn, 241–242
secondary neutrons, 137–138, 209, 212–213, 225, 227, 243–244
Second Industrial Revolution, 6–7, 12–13, 22, 26

Index

Second World War
  German invasion of Poland, 218
  heavy water movement, 227–228
  Japan's surrender, 268
  Nazi persecution of nonethnic Germans, 142
  Nuclear Age and, 7
  Pearl Harbor, 238
  sonar, 70
Segrè, Emilio
  Fermi's theory, 155
  inducing radioactivity, 169
  on liberation of nuclear energy, 166
  on Majorana, 101
  on Noddack, 174, 176101
  photograph, 181 (fig.)
  plutonium bomb, 254
  Rome laboratory, 96–97
shake (unit of measurement), 268
shells, energy, 133–134
shockwave
  in bomb detonation, 256, 274–276, 287
  from Joe-4 bomb, 286
  from Trinity test explosion (1945), 6, 91
  from Tsar Bomba, 297–298
silver, irradiation of, 178
Sloika, 284
slow neutrons, 178–180, 192–193, 210, 216, 219, 225, 236
Snow, C. P., 110, 223–224
Soddy, Frederick
  afterword on, 312
  Interpretation of Radium (Soddy), 40
  isotopes, 49–50, 52, 62
  radioactivity, 36–37, 39–42, 49–50
  vision of nuclear energy, 299
Solvay, Ernest, 142
Solvay Conference (1933), 141–146, 141 (fig.), 152–154
sonar, 69–70
sound
  speed of, 5, 110
  Trinity test explosion (1945), 5–6
Soviet Union
  fission experiments, 229–230
  fusion bomb, 280–288
  German invasion, 230
  nuclear weapons, 259–260
  spies, 259–260, 280, 282–284, 289, 291, 308
  Tsar Bomba, 293–298, 303–304, 309
spies, 259–260, 280, 283–284, 289, 291, 308
splitting atoms
  barium from splitting uranium, 196, 199, 201
  beryllium, 138, 138 (fig.)
  boron, 124
  Cockcroft and Walton, 121–124
  lithium, 121–123, 134–138, 138 (fig.)

Rutherford, 69–80
thorium, 214–216
uranium by Irène Joliot-Curie, 208–209
Stalin, Joseph, 230, 260, 285
Stalin Prize, 287, 303, 309
statistical mechanics, 94–95
steam power, 6–7, 11
Strassmann, Fritz, 193, 195–196, 305
sun, fusion in, 262–264, 302
Sundial, 304
Super Bomb, 264–266, 270, 274–275, 283, 289, 302. *See also* Classical Super bomb
supercritical, 3, 232, 254, 261, 271
surface tension, 193, 198, 200–201
Szilard, Leo, 135–139
  afterword on, 313
  atomic bomb, 224
  chain reaction, 136–139, 189, 224
  fission, 212–213
  petition against bomb use, 258

Tamm, Igor, 282–285, 289, 302
tantalum, 12
Taylor, Brian, 290
technetium, 176
Teller, Edward
  afterword on, 314
  Alarm Clock, 274, 276, 284
  Fermi and, 262–264, 269
  at Los Alamos, 265–266
  personal characteristics, 262–263
  pursuit of big bombs, 303–304
  Szilard meeting, 212
thermodynamics
  Fermi, 93
  kinetic energy of heat, 11
  laws of, 30, 53
thermonuclear weapons, power of, 261–262
Third Industrial Revolution, 6
Thomson, Joseph John (J. J.)
  afterword on, 311
  Bohr and, 60
  cathode rays, 22–23
  electron discovery, 23–24, 28, 31, 36, 54
  Nobel Prize, 24
  Rutherford and, 24, 27–29, 31
thorium
  Curie experiments, 20
  emanation, 34–35, 37
  Hahn experiments, 43–47
  isotopes, 43–47
  Rutherford experiments, 30–31, 33–38
  splitting of, 214–216
Three Mile Island disaster, 301
The Time Machine (Wells), 188
tokamak, 302–303

Index

"Tom, Dick, and Harry," 290
transmutation
  by alpha radioactivity, 50
  Bohr's explanation of, 62
  Frédéric Joliot on explosive, 166
  frozen energy, 41
  nitrogen to hydrogen, 76
  nitrogen to oxygen, 74, 75 (fig.)
  radiation as direct result of, 50
  Sakharov, 282
  spontaneous, 37, 76
  thorium to radon, 37
transuranic elements
  Fermi, 171–173, 177, 196, 213, 239
  Meitner, Lise, 193–194
  production of, 215
Trinity (1945), 1–6, 91–92, 257, 279
tritium
  generation within bomb, 284–285
  heavy hydrogen, 78, 133
  in hydrogen bomb, 262, 266–272, 274–275, 283–285, 290, 302
  production inside bomb, 274–275, 275 (fig.), 284–285
triton, 78, 266, 267 (fig.)
Tsar Bomba, 293–298, 303–304, 309
Tube Alloys, 235, 249, 259
Tuck, James, 256–257

Ulam, Stanislaw, 273–277, 314
Ulam–Teller invention, 277, 279–281, 287–291
uncertainty principle, 131, 133, 151, 153
United Kingdom
  atom bomb research shared with U.S., 236–238
  hydrogen bomb development, 288–291
uranium
  Becquerel's experiments, 17–20, 18 (fig.)
  Curie experiments, 20–21
  decay to lead, 44
  fission, 197–105, 208–209
  half-life, 171
  as heaviest element found on Earth, 3, 12, 50, 170, 185, 189
  number of neutrons, 185
  in periodic table, 12
  plutonium conversion to, 3
  radioactive ladders, 51 (fig.)
  Rutherford experiments, 29–33, 38

uranium-235
  discovery, 217
  enrichment, 224–225, 230–232, 234, 237, 249, 251–252
Uranium Committee, 224–225, 237, 262

vacuum pump, 14, 23
Via Panisperna boys, 97–98
volcanic eruption, 294
von Neumann, John
  afterword on, 314
  boosted fission bomb, 270–271
  Bronowski on, 255, 269, 271
  computer science, 269–271, 275–276, 278
  early life and career, 255
  Einstein on, 255
  Fuchs–von Neumann device, 272, 291
  hydrogen bomb, 269–272
  ionization compression, 270
  at Los Alamos, 255–257, 269–275
  as Ulam fellow traveler, 273
von Weizsäcker, Carl, 186, 200

Walker, James, 35
Walton, Ernest, 116–117, 120–124, 134, 137, 313
Ward, John, 289–291
Watt, James, 6
wavelength, 65
Webster, Hugh, 104–105
Weil, George, 245–247
Wells, H. G., 40–41, 106, 188
Wick, Gian-Carlo, 98, 183, 185
Widerøe, Rolf, 117–118
The World Set Free (Wells), 40–41, 188, 248

X-rays
  Curie, 86
  energy measurement, 61
  in fusion bomb, 277–279
  Meitner, 49
  Röntgen's discovery, 16, 19, 28
  Rutherford, 28–29, 32

York, Herbert, 279–280

Zeldovich, Yakov
  afterword on, 315
  chain reaction, 230
  Soviet atomic bomb project, 282–283, 285, 287
  uranium-235 enrichment, 232

Credit: PEW Library

FRANK CLOSE, OBE, FRS, is a particle physicist and an emeritus professor of physics at the University of Oxford. He is the author of over two dozen books, including *Elusive, Half-Life,* and *The Infinity Puzzle.* He lives near Oxford, UK.